Engineering the Climate

Engineering the Climate

The Ethics of Solar Radiation Management

Edited by Christopher J. Preston

LEXINGTON BOOKS
Lanham • Boulder • New York • Toronto • Plymouth, UK

Published by Lexington Books
A wholly owned subsidiary of Rowman & Littlefield
4501 Forbes Boulevard, Suite 200, Lanham, Maryland 20706
www.rowman.com

10 Thornbury Road, Plymouth PL6 7PP, United Kingdom

Copyright © 2012 by Lexington Books
First paperback edition 2014

All rights reserved. No part of this book may be reproduced in any form or by any electronic or mechanical means, including information storage and retrieval systems, without written permission from the publisher, except by a reviewer who may quote passages in a review.

British Library Cataloguing in Publication Information Available

Library of Congress Cataloging-in-Publication Data

The hardback edition of this book was previously cataloged by the Library of Congress as follows:

Preston, Cristopher J. (Christopher James), 1968-
 Engineering the climate : the ethics of solar radiation management/ Christopher J. Preston
   p. cm.
Includes bibliographic references and index.
1. Solar radiation—Management. 2. Solar radiation—Moral and ethical aspects. 3. Climate change. I. Title.
 QC911.P747 2012
 174'.93337923—dc23
                    2012012642

ISBN 978-0-7391-7540-8 (cloth : alk. paper)
ISBN 978-0-7391-9054-8 (paperback)
ISBN 978-0-7391-7541-5 (electronic)

∞™ The paper used in this publication meets the minimum requirements of American National Standard for Information Sciences—Permanence of Paper for Printed Library Materials, ANSI/NISO Z39.48-1992.

Printed in the United States of America

Contents

Acknowledgments	vii
Introduction	ix
The Extraordinary Ethics of Solar Radiation Management *Christopher J. Preston*	1
Present and Future Generations	13
1 Geoengineering, Solidarity, and Moral Risk *Marion Hourdequin*	15
2 Might Solar Radiation Management Constitute a Dilemma? *Konrad Ott*	33
3 Domination and the Ethics of Solar Radiation Management *Patrick Taylor Smith*	43
Marginalized, Vulnerable, and Voiceless Populations	63
4 Indigenous Peoples, Solar Radiation Management, and Consent *Kyle Powys Whyte*	65
5 Solar Radiation Management and Vulnerable Populations: The Moral Deficit and its Prospects *Christopher J. Preston*	77
6 Solar Radiation Management and Nonhuman Species *Ronald L. Sandler*	95
Moral Hazards and Hidden Benefits	111

7 The World That Would Have Been: Moral Hazard Arguments
   Against Geoengineering   113
   Ben Hale

8 Climate Remediation to Address Social Development
   Challenges: Going Beyond Cost-Benefit and Risk Approaches
   to Assessing Solar Radiation Management   133
   Holly Jean Buck

Ethics of Framing and Rhetoric   149

9 Insurance Policy or Technological Fix? The Ethical Implications
   of Framing Solar Radiation Management   151
   Dane Scott

10 Public Concerns about the Ethics of Solar Radiation Management   169
   Wylie Carr, Ashley Mercer, and Clare Palmer

The Cultural Milieu   187

11 The Setting of the Scene: Technological Fixes and the Design of
   the Good Life   189
   Albert Borgmann

12 Between Babel and Pelagius: Religion, Theology, and
   Geoengineering   201
   Forrest Clingerman

13 Making Climates: Solar Radiation Management and the Ethics
   of Fabrication   221
   Maialen Galarraga and Bronislaw Szerszynski

Bibliography   237

Index   255

Contributors   263

Acknowledgments

Research time for writing this book was made possible by a grant from the National Science Foundation (SES # 0958095). Invaluable help in many of the editorial stages was provided by Andrea Gammon (whose research assistance was made possible by the same National Science Foundation grant). Andrea's helpful demeanor and incredibly hard work made this process much more straightforward than it might otherwise have been. I am extremely grateful to her. I would like to thank Holly Jean Buck who volunteered her proofreading and copyediting skills at an opportune moment. I would also like to express gratitude to the other members of the grant team at the University of Montana, Dane Scott, Laurie Yung, and Wylie Carr for providing a professional and collegial working atmosphere, and to the University of Montana and the wider Missoula community for providing an exceptional environment in which to focus on these important issues. Finally, thanks—as always—to Lisa and my family for their unconditional support and love.

Introduction

The Extraordinary Ethics of Solar Radiation Management

Christopher J. Preston

In his 1989 book, *The End of Nature*, Bill McKibben made a powerful case that there is something of great philosophical significance to climate change. He argued that our impact on the atmosphere through the burning of fossil fuels has pushed the human footprint on the earth over an important ethical threshold. After Al Gore's documentary, *An Inconvenient Truth* (2006), increasing numbers of people have been willing to see the passing of this threshold—and the need to do something about it—as a pressing moral, as well as a difficult scientific, issue. Large segments of the public have come to realize that anthropogenic warming creates unique philosophical and ethical challenges. Stephen Gardiner recently labeled these challenges "the perfect moral storm" (2011a) and has explained in helpful detail why they are challenges we seem particularly ill-prepared to address.[1]

The idea of deliberately manipulating the climate system in order to push back against global warming takes humanity into moral terrain that is, if anything, even more challenging than the terrain entered through inadvertent climate change. The suite of emerging technologies known as climate engineering (or geoengineering) brings a new twist to the philosophical and moral issues raised by McKibben, Gardiner, and others. No previous technology has intentionally manipulated the earth at such a fundamental level, deliberately altering a system with so much complexity and promising such widespread effects. If initiated, climate engineering would take anthropogenic influence on the earth to a whole new level. It would mean intentionally assuming responsibility for the very skies under which all life on earth lives, an endeavor with repercussions impacting everyone—and everything—on the planet.

Since the beginning of contemporary research into climate engineering there has been a wide acknowledgment that the ethical issues match the scientific ones for their complexity. On the one hand, geoengineering, if successful, could massively reduce the amount human suffering and environmental loss predicted to be in store as a result of global warming. On the other, geoengineering raises significant new ethical concerns of its own. In one of the first articles in the scientific literature on geoengineering, environmental scientist Stephen Schneider and atmospheric scientist William Kellogg pointed out that the inherent uncertainty about the effects of the technology would create a moral obligation to compensate any potential "losers" resulting from intentional climate modification (1974). Broadening the ethical purview at a 1994 American Association for the Advancement of Science (AAAS) symposium, environmental scientist Gregg Marland admitted that, with geoengineering, it is "hard to focus on the 'could we' parts of the discussion without raising the 'should we'" (1996, p. 276). In "Albedo Enhancement by Stratospheric Sulfur Injections," the essay credited for turning intentional management of solar radiation into a serious public policy option, Nobel prize-winning chemist Paul Crutzen conceded that, despite the promise of stratospheric aerosols for countering global warming, the social and ethical issues "are many" (2006, p. 217). The UK's Royal Society captured all these sentiments by claiming in its 2009 report that "the greatest challenges to the successful deployment of geoengineering may be the social, ethical, legal and political issues associated with governance, rather than scientific and technical issues" (2009, p. xi). This recognition of the hazardous (and perhaps determinative) nature of the moral terrain even amongst participating scientists is one of the unusual things about the nascent geoengineering discourse.[2] Given this awareness of the gravity of the ethics, it seems quite likely that climate engineering may be one of those few technologies where the ethical considerations could determine whether the technology ever gets implemented at all.[3] A comprehensive evaluation of the ethical issues it raises is therefore an essential part of any concerted research program.

Climate engineering is today separated out into two main types.[4] Carbon dioxide removal (CDR) technologies extract carbon dioxide from the atmosphere in order to directly address one of the main drivers of global warming. CDR strategies include large scale afforestation and reforestation projects, the deliberate proliferation of phytoplankton by supplementing ocean environments with iron or other nutrients, enhancing the weatherization of rocks, the large-scale pyrolysis of terrestrial vegetation, increasing the alkalinity of the seas, and the deployment of many thousands of artificial trees. All of these technological methods would remove carbon dioxide from the atmosphere with the intention of safely sequestering it somewhere benign for the long term. If carbon dioxide is considered a pollutant, CDR can be viewed simply as a type of pollution control. CDR technologies have a number of considerations in their favor, including public familiarity with the idea of pollution control, the fact that they directly attack the cause of the warming problem, and the fact that most of them are scalable. The more similar to traditional pollution control a particular CDR technology looks, the more support it tends to receive, even if a number of the technologies turn out upon closer scrutiny to come with their own environmental concerns (Trick, 2010; Garcia Perez, 2008).

Solar Radiation Management (SRM) technologies—also known as sunlight reflection methods—seek to increase the reflectivity of certain planetary boundaries in order to lessen the solar energy reaching the earth and its atmosphere. SRM possibilities include putting mirrors in space, increasing the amount of aerosols in the stratosphere, brightening existing marine clouds, and whitening large areas of terrestrial surfaces. Because SRM leaves carbon dioxide in the atmosphere and, in its atmospheric forms, involves "unencapsulated"[5] technologies, it is widely regarded as the more controversial of the two types of geoengineering. Because the space-based, stratospheric, and (to a lesser extent) tropospheric manipulations of earth's albedo envisioned in SRM are not as easily scalable as CDR, the moral issues they raise tend to become global more quickly than with CDR. The fact that SRM is directed at something as fundamental as the amount of energy the earth receives from the sun means that these technologies initially give the appearance of impacting Nature (writ large) in a way that CDR technologies do not.[6]

Despite these negative factors, SRM technologies in general—and stratospheric aerosols in particular—have the advantage, as David Keith points out, of being "fast" and "cheap" (Keith et al., 2010, p. 426).[7] If geoengineering ever had to be deployed in some sort of critical planetary emergency as an option of last resort—a circumstance some suggest is the *only* circumstance in which geoengineering should ever be deployed—it looks today as if stratospheric aerosols would be the chosen treatment. SRM, particularly SRM through the deployment of stratospheric aerosols, therefore deserves special scrutiny. It provides a unique and acute set of issues for ethicists to contemplate. The essays in this volume are addressed toward unpacking some of the most challenging of them.

INITIAL SKETCHES OF THE ETHICAL TERRAIN

As the quote from the Royal Society made clear, it is the governance questions that initially appear to be the greatest obstacles to SRM. Questions about how to involve stakeholders in decision making, how to mitigate the different risks to various populations caused by SRM's uncertain effects, and how to ensure the legitimacy of any political body that might initiate SRM deployment have no easy answers. The Solar Radiation Management Governance Initiative, a joint effort initiated by the Royal Society, the Academy of Sciences for the Developing World, and the Environmental Defense Fund, has brought together a group of partners to generate guidelines for how to answer some of these questions.[8] While governance questions demand legal and policy answers, the root of these challenges is fundamentally ethical. In fact, the reason why the prospect of governing SRM is so daunting is that the ethical criteria one would hope to meet in governance are so numerous and so difficult.

Adam Corner and Nick Pidgeon (2010) have helpfully identified a number of general areas of ethical concern raised by geoengineering. One of the first areas they identify is the issue of consent. How would those initiating any SRM project ensure the consent of those likely to experience the effects of an engineered climate? All citizens, they point out, could claim "to have a legitimate stake . . ." in the " . . . control of the global thermostat" (Corner & Pidgeon, 2010, p. 29). While it is not typically thought that every citizen on earth must consent to a governance policy before it can be implemented, difficult questions concerning procedure and representation clearly arise with climate engineering given the widespread and uncertain nature of its impacts. Questions about consent are made more complex by the fact that decisions on geoengineering today—even if they are decisions *not to* geoengineer— will have a large impact on the lives of future generations (as well as on the question of who will actually exist in those generations[9]). Adding to the ethical challenges, those countries with the most to lose in the face of runaway climate change (and therefore the most to gain should SRM successfully avert the projected harms) are often those with the least political power. These include countries likely to suffer as a result of sea level rise (e.g., Bangladesh) and countries likely to be most harmed by increasing drought conditions (e.g., countries in sub-Saharan Africa). The moral quandaries are made even more acute by the fact that the countries facing the most harm are typically the ones that have contributed least to creating the problem of global warming in the first place due to their underdevelopment. Issues of consent, participation, and representation therefore loom large among the moral challenges of SRM.

Another category of moral challenge Corner and Pidgeon identify is the question of whether there might simply be something wrong with the whole idea of manipulating something as large and as complex as the global climate. Both "extrinsic" and "intrinsic" arguments can be offered for this position.[10] The extrinsic line of thinking suggests that the complexity of the climate system ensures there will be numerous unanticipated harms stemming from climate engineering. Current modeling suggests plenty of uncertainty about SRM's regional effects on temperature, precipitation, and the hydrological cycle. In some circumstances, it is conceivable that SRM could create harms that exceed the harms of global warming itself.

The intrinsic arguments point not toward possible harms but suggest instead that it may be wrong on principle to meddle intentionally with something as fundamental as solar radiation. Doing so appears to be inconsistent with the principles of many types of environmentalism (Preston, 2011) and might display an inappropriate hubris about the proper role of humans in the biosphere (Jamieson, 1996). As Clive Hamilton has put this latter point, "what matters ethically about geoengineering is not only the outcome but also the human disposition it reveals" (2011c, p. 18).

The third category of concern identified by Corner and Pidgeon involves global security and law. Though the United Nations Treaty on Environmental Modification (1977) banned the modification of the climate for hostile purposes, there is a worry that some nations might seek strategic advantage through certain types of climate modification. The fact that the deployment of stratospheric aerosols is cheap enough and technically simple enough to be pursued unilaterally (even, perhaps, by a particularly wealthy individual or corporation) means that the potential for politically destabilizing unilateral climate engineering is real. Military, economic, and even terrorist motivations are all possible. The task of finding appropriate legal mechanisms to lessen these possibilities is one of the biggest challenges faced by the international community. In the event of any globally legitimate SRM regime, considerable geopolitical stability would be required in order to maintain the necessary management until greenhouse gases could be adequately reduced and SRM withdrawn. Ensuring this stability in a climate-stressed world would not be easy.

"The lure of the techno-fix" creates a fourth category of ethical concern. One of these lures is that if temperatures could be reduced with engineering techniques, this may dissuade nations from making the necessary reductions in carbon dioxide emissions. This problem, known as the "moral hazard" of geoengineering, suggests that geoengineering encourages irresponsible behavior that would be less likely in the absence of a technical fix. A more general version of this worry is that a technical solution to the greenhouse gas problem is really the wrong response to what is essentially a social problem involving overconsumption. The necessary move to a largely post-carbon economy and the (arguably) desirable move to a less consumptive society might be hindered by this lure of the techno-fix. A different type of problem created by the potential techno-fix for global warming is the possibility of commercial interests taking over the task of cooling the planet, exposing everyone to the danger that the interests of those with the deepest pockets will prevail.

In addition to the categories identified by Corner and Pidgeon, there have been several other notable contributions to discussions of the ethics of SRM. In the first article by a professional philosopher on the ethics of intentional climate change, Dale Jamieson (1996) reflected on the intrinsic argument against climate engineering before discussing four conditions that any project of intentional climate modification must meet. It must: a) be technically feasible; b) have predictable consequences; c) produce economic states preferable to the alternatives; and d) not violate any well-founded ethical principles. Echoing several of the concerns already mentioned, Jamieson argued that these ethical principles include adherence to democratic decision making, avoidance of irreversible change, and learning to live with nature. At the time he wrote the article—and arguably still today—no geoengineering deployment could satisfy the conditions Jamieson laid out.

Expressing concern about SRM on the grounds that "the cure might be worse than the disease," atmospheric scientist Alan Robock (2008, p. 14) provided twenty reasons why geoengineering might be a bad idea. Several of these reasons introduced additional ethical concerns. Robock mentioned the aesthetic cost of a blue sky potentially whitened by stratospheric aerosols, a possible reduction in the generating potential of photovoltaic panels, the effects on plant productivity of more diffuse light, and negative impacts on astronomy and satellite imaging. Each of these consequences raises additional ethical concerns for those with specific interests. Robock added that the inability to extract aerosols from the stratosphere once they had been deployed raised further concerns about making an irreversible commitment to a path involving significant future risk.

Because actual management of solar radiation probably lies some years into the future, at least one ethicist has focused not on the ethics of geoengineering itself but on the ethics of initiating geoengineering research. By developing arguments from his analysis of climate change as the perfect moral storm, Stephen Gardiner (2010) has shown how the decision to embark on geoengineering research potentially manifests various forms of moral corruption. The self-interest that the current generation of wealthy consumers has in postponing any meaningful action on climate change makes modest research into geoengineering a deceptively attractive option. Gardiner suggests that the portrayal of geoengineering as the "lesser of two evils" and the temptation to begin research on it today under the guise that it is a legitimate and responsible component of climate policy might instead be a demonstration of "the subversion of our moral discourse to our own ends" (Gardiner 2010, p. 286). Geoengineering may not in the end be wrong. There is, however, a high potential for our reasoning about it to be distorted. The interests of the powerful in the current generation could easily take precedence over those of the poor in the future. If Gardiner is right about the potential for moral corruption, it may not only be the practice of geoengineering that creates moral issues. Even directing attention to it as a legitimate strategy comes with a moral cost, a cost that might be enough to blight humanity far into the future (Gardiner, 2010, p. 304).

SURVEY OF CONTENTS

The current collection pursues some of the themes mentioned in further depth and introduces a number of new ones. Three of the chapters in this anthology illustrate the different ways that SRM might morally impact current and future generations. Each contribution shows how the decision to start down the path of SRM creates extraordinary conditions both for today and for posterity. Marion Hourdequin points out that SRM undermines first the intragenerational and then the intergenerational solidarity needed to make the just decisions required to really address the problem of climate change. Due to its one-sided causes and projected differential impacts, SRM could seriously compromise solidarity among present, and between present and future, publics.

Konrad Ott is concerned about the slippery slope down which today's SRM discourse unwittingly takes us. He highlights the moral hazard associated with embarking on SRM research, suggesting there is a "performative force" to SRM discourse that will lead to the neglect of other strategies for dealing with climate change. One consequence of this neglect is that future decision makers will be forced into the dilemma of having to choose between continuing SRM (even if it becomes dangerous) and terminating SRM (with the consequence of extremely rapid warming). Ott concludes with a recommendation for a moratorium on SRM field tests until at least 2040.

Continuing the worries about its impact on the future, Patrick Taylor Smith shows that SRM engenders a particularly pernicious type of domination of future generations. Having carefully laid out why domination is a more suitable lens through which to view concerns about future generations than harm, Smith argues that, because SRM merely blocks (rather than eliminates) a serious threat, it necessarily imposes limiting conditions on the choices of posterity. As much as saving future generations, SRM coerces them into a particular set of choices while letting the current generation largely off the hook for its follies.

In the next section of the collection, the concerns of the traditionally marginalized and/or the voiceless are considered. Kyle Powys Whyte writes the first article from an indigenous perspective on the ethics of geoengineering. He considers how tribal sovereignty adds a unique challenge to the question of informed consent. Whyte argues through examples that the United States possesses a poor track record for involving indigenous people in early discussions concerning technologies that will affect them. He details three models for securing tribal participation in decision making, suggesting that only the third, the "partnership model," contains any promise for adequate tribal involvement.

Making use of Henry Shue's (1992) notion of "compound injustice," Christopher J. Preston suggests that the widely acknowledged moral deficit owed by the rich nations to the poor over climate change has the potential to be dramatically compounded by SRM. In order to stave off this potential worsening of injustice, Preston offers the notion of "desert" as showing some promise for keeping the door open to a wide enough range of forms of redress. He discusses the implications of this deficit on current plans for SRM research and suggests that, at the very least, the poor nations deserve early input into the form of SRM technologies.

Ronald Sandler considers our moral obligations to a population that will always remain voiceless in climate negotiations: the nonhuman world. First Sandler makes the case for species having "final value" and shows how that value is currently threatened by climate change. Then he suggests that SRM's inherent uncertainty and its failure to address the presence of carbon dioxide in the atmosphere mean that it is unlikely to be a desirable option from the point of view of the final value of nonhuman species. At the end of the chapter, however, he leaves the door open to the possibility that there might be circumstances under which SRM will not make the situation of nonhuman species any worse than it will already be under climate change.

One of the most commonly cited worries about the prospect of engineering our way out of the climate problem is that it will remove some of the incentive for reducing emissions. Ben Hale helpfully unpacks the ambiguity in a number of versions of this "moral hazard" claim. After identifying six main strands of the moral hazard, Hale examines sixteen often heard complaints about geoengineering through these strands. In each case, it turns out more work needs to be done to show there is an actual wrong involved. Without this work, the label of the moral hazard serves only as "a falsely concrete straw man."

Holly Jean Buck suggests that SRM may not present a moral hazard at all but a moral opportunity. "Not all SRM strategies," claims Buck, "are created equal." Offering an alternative narrative, Buck defends the surprising claim that certain types of SRM, if done properly, have socially transformative co-benefits to them. The beginning of a solution to some trenchant social problems might emerge out of particular types of climate engineering, if they are embarked upon carefully and with considerations of justice at the forefront.

As a relatively new technology, geoengineering is currently creating massive interest in print media and the blogosphere, but the way the technology gets portrayed is still very much up for grabs. Dane Scott looks at the different ways SRM is currently being framed. He examines in detail two operative frames for SRM in common use: "Insurance/Plan B" and "Technological fix." He carefully demonstrates how the different frames highlight different issues, some of which can be misleading and unhelpful. By focusing on one particular framing, the "pragmatic technical fix," Scott hopes to identify a set of issues that need much more attention than they currently receive.

In their contribution to the volume, Wylie Carr, Ashley Mercer, and Clare Palmer show how recent social science data on geoengineering reveal that the public are already apt to view SRM in fundamentally *moral* terms as much as scientific or technical ones. Public perceptions, however, can express a different vision of a technology from the vision of scientists and policymakers, thereby raising different moral issues. Carr, Mercer, and Palmer sift through some initial public responses to geoengineering and highlight the dominant ethical concerns. These social science results, the authors argue, can help steer the discussion of the ethics of SRM in directions that specialists might not have anticipated.

The final section of the book pays attention to how the discussion of the ethics of SRM always takes place within a wider cultural milieu. Philosopher of technology and society Albert Borgmann examines how our decisions in matters of large significance such as climate change gain credence from the cultural setting in which they are made. SRM has been presented within a technocratic cultural setting that "defuses" the need for us to make bigger decisions about our lifestyles. While remaining non-committal on SRM itself, Borgmann asks us to consider who we should let determine the broader contours of our cultural lives and what kind of vision of the good life that setting should support.

Forrest Clingerman's contribution recognizes the important role of religion in shaping the kind of interactions with the cosmos that are deemed appropriate or profane. Drawing on resources from Christian theology, Clingerman explores SRM through a contrast between the hubris implicit in the Tower of Babel and the careful assumption of responsibility found within Pelagianism. The contrast points toward a balance between fallibility and capability that, Clingerman suggests, any theological response to SRM must strike.

Maialen Galarraga and Bronislaw Szerszynski scrutinize what they call "making climate" through SRM. They identify three different "modes of making," each of which says different things about what is expected of the maker and what is understood about her relationship to the environment. Galarraga and Szerszynski claim there are important moral differences between the roles of architect, artisan, and artist, and they find the latter matches most closely the role of those who might "make climate" through SRM.

Taken together, these essays expand the range of moral issues deemed pertinent to SRM and add considerable depth to the discussion of several of them. Bad arguments are exposed, new issues are raised. The essays often provide warnings about commencing SRM, but in the process they add a great deal of clarity to the question of the conditions that must be satisfied if geoengineering is to proceed ethically. Certainly the essays in this volume do not exhaust the list of ethical issues that climate engineering raises. For example, important considerations concerning SRM's impact on the value of wild and unmodified nature still linger in the background. Worries about climate "hacking" for terrorist or other coercive purposes also demand further attention. Discussion of how SRM might fit within the purview of environmental restoration must still be pursued. What these essays do offer is a generous sampling of views on what is most morally significant about the management of solar radiation. These are early days and the ethical landscape is only just beginning to come into focus. If researchers in climate engineering are going to make good on the promises of the early practitioners to take the ethical challenges seriously, then the issues discussed here should be given more than a casual nod of passing acknowledgment. They should be viewed as an essential part of the ongoing research context for geoengineering as a whole.

NOTES

1. Gardiner's book offers a detailed explanation of the spatial, intergenerational, and theoretical dimensions of these challenges.

2. This discourse warrants the label "nascent" because, at time of writing, very little research money is being devoted to geoengineering at a national government level. In 2010, the U.S. Government Accountability Office review of geoengineering activities found that "due to limited federal investment, it is premature to coordinate geoengineering activities" (U.S. GAO, 2010 (highlights)). In the UK, the Engineering and Physical Sciences Research Council devoted only £3 million to climate engineering in 2010. A small amount of private funding is also available. Bill Gates, for example, has personally endowed a Fund for Innovative Climate and Energy Research administered by David Keith and Ken Caldeira that supports, in part, geoengineering research.

3. Human cloning is one of the rare examples of a technology in which the ethical arguments have proven decisive against its introduction.

4. This distinction has been widely adopted since Royal Society (2009).

5. The U.S. Congressional Research Service Report on geoengineering distinguishes between encapsulated and unencapsulated technologies, suggesting that the former tend to be viewed as more ethically acceptable. The distinction revolves around ". . . whether a technology program is modular and contained or whether it involves the release of materials into the wider environment" (Bracmort et al., 2010, p. 5).

6. Stephen Gardiner (2011b) questions these first reactions about the relative acceptability of CDR over SRM.

7. Keith adds that these advantages are qualified by the disadvantage of it also being "imperfect."

8. Details are available at www.srmgi.org.

9. These are problems originally introduced to ethics by Parfit (1984).
10. This language was employed in the GMO debates to distinguish between the idea that genetically modifying species might lead to human (or environmental) harm (extrinsic) and the idea that manipulating genes is wrong in itself regardless of the consequences (intrinsic). See, for example, Comstock (2010).

Present and Future Generations

Chapter One

Geoengineering, Solidarity, and Moral Risk[1]

Marion Hourdequin

> Climate change is different from other problems facing humanity—and it challenges us to think differently at many levels. Above all, it challenges us to think about what it means to live as part of an ecologically interdependent human community.
>
> —(United Nations Development Program, 2007, p. 2)

Anthropogenic global climate change is frequently characterized as a collective action problem, or more specifically, as a tragedy of the commons (Gardiner, 2001; Johnson, 2003). In a tragedy of the commons, the desired collective good can be secured only through the combined action of many individuals and institutions, yet each individual (or institution or nation-state) lacks the incentive to sacrifice their individual interests to secure the collective good. Without the assurance that others will act to secure the general good, the "rational" thing to do seems to be to act in one's own narrow interests. However, when each individual acts in this way, tragedy ensues: the commons is depleted, polluted, damaged, or destroyed.

Collective action problems pose a very real threat to the global environment, and particularly, to a stable and hospitable climate. Finding a way to solve such problems is thus a formidable yet critical challenge. Fortunately, individuals do sometimes set aside their own narrow interests to contribute to the provision of a collective good, and such behavior can generate similar responses in others (see, e.g., Fischbacher, Gächter, & Fehr, 2001; Fehr & Fischbacher, 2004). In experimental public goods games, people consistently contribute to the public good, and contributions increase when participants have the opportunity to see and talk with other players (Ostrom, 2000). It seems that even without top-down controls, certain conditions can disrupt the tragedy of the commons. In particular, a stronger sense of moral community seems to help avert tragedy: communities that are successful in solving collective action problems tend to have a high degree of trust, reciprocity, and fairness, and they adhere to an established set of social norms (Ostrom, 2000).

Recent research on collective action problems delivers two key messages. First, people do not act only to maximize material benefits to themselves; they also act on social and ethical norms (Ostrom, 2010, p. 156). Second, *context matters*: certain circumstances support individual choices and institutional arrangements that promote the broader public good; others do not. These points have important implications for the ethics of climate change. The first suggests that in considering what we ought to do, as individuals or collectives, we need not assume a world of rational maximizers. This is encouraging. Yet we cannot assume a world of altruists either; most people's motives are mixed. Some lean toward altruism, while others act more consistently from a self-interested point of view. The second point, emphasizing context, helps us see how we might respond to this complex mixture of motives. Because context shapes the possibility of motivation by moral norms, we need to determine not only how to address collective action problems directly but also *how to produce the conditions that make possible actions to address such problems.*

This means that if we are concerned with a just solution to the problem of global climate change, we should be concerned not only with the *direct* effects of various approaches to addressing the problem but also with their *indirect* effects. More specifically, we should be concerned not only with how climate strategies affect the physical environment but with how they affect the moral environment. That is, we need to be concerned with the conditions that enable the establishment of a just climate regime.

The aim of this paper is to consider how certain kinds of geoengineering may influence the moral climate (see Gardiner, 2010, especially section VI, and Gardiner, 2006b, for broader discussion of the relation between climate change and the moral climate), and in so doing, enable or inhibit an ethical solution to the challenge of global climate change. More specifically, I argue that large-scale solar radiation management (SRM)—which aims to combat global warming by reflecting greater quantities of solar radiation back into space—has the potential to make more difficult the establishment of just climate policies and practices. I claim that even if SRM techniques could secure certain benefits, they pose serious risks to *moral solidarity*, and as such are likely to make a just, long-term solution to climate change more difficult than it already is.[2]

In the next section of the paper, I provide some brief background on climate ethics and climate justice. I then turn to a discussion of solidarity and its relevance to geoengineering and global climate change. Here I explain how geoengineering threatens solidarity, both internationally and intergenerationally. I also connect the solidarity argument to a more general discussion of our obligation to consider threats to the moral climate.

CLIMATE ETHICS AND CLIMATE JUSTICE

Before turning to the question of solidarity, let me briefly review some key issues in climate ethics. Climate change raises concerns about *harms* and their *distribution*. As is well known, global warming is not only raising the earth's global mean surface temperature, it is causing glacial retreat, melting of polar ice, sea level rise, and changes in patterns of precipitation, seasonal snowmelt, river flows, hurricanes, and other severe storms (IPCC, 2007b; IPCC, 2007c; Karl et al., 2008). The distribution and abundance of plants and animals throughout the world is changing as species migrate toward the poles and upward in elevation, and extinction rates are expected to rise significantly as the climate continues to warm (Walther et al., 2002). Scientists also predict shifts in human population distribution as low-lying areas, including island nations, are flooded and other areas become less habitable due to drought, desertification, and other climate-related issues (Warner et al., 2009). More people are likely to suffer as a result of extreme weather events, and those in less developed countries, especially, may suffer malnutrition as a result of climate-related crop failure (IPCC, 2007c). Climate change also may destabilize international geopolitics as resource scarcity increases (Barnett & Adger, 2007). Clearly, climate change will result in significant harms to individuals, nations, plants, animals, and ecosystems, and these harms are of ethical concern.

Furthermore, it is well established that the impacts of climate change will not be borne equally around the world (IPCC, 2007c). Many of the nations that contributed least to historical greenhouse gas emissions and continue to contribute proportionately little today will be affected most strongly by sea level rise, changing precipitation patterns, intensification of storms, and other climate-related phenomena. The irony of this situation is further compounded by the fact that these nations are, in general, poor and not well positioned to adapt to climate change through infrastructural or other adjustments that would moderate its effects. Nor are they likely to be the leaders and primary decision makers in geoengineering schemes. Already the inequities in the distribution of climate harm have become a significant source of international disagreement.

Adequately addressing the problem of climate change thus requires both *minimizing harms* and working toward a *fair distribution of climate burdens*. A fair distribution requires attention both to the distribution of actual climate harms and to the distribution of the costs (financial and otherwise) of addressing the climate problem. While there is no way to evenly distribute the impacts of climate change across the globe, inequities in the actual distribution of climate harms could be mitigated by reducing the total amount of harm produced through the prevention of serious climate change and through assistance to those most strongly impacted by climate change.

Although there is debate over the precise structure of a just climate regime, there is substantial convergence among various philosophical views on the basic distribution of responsibility in addressing climate change. This convergence results because many of the most plausible moral principles for allocating responsibility each point in the same direction (Shue, 2010). For example, a principle that allocates responsibility among nations based on past contributions to the problem (i.e., past greenhouse gas emissions) points in the same direction as a principle that allocates responsibility based on ability to pay. Similarly, a principle that recommends particular attention to the least well off or that recommends guaranteeing each individual a decent minimum, will recommend a distribution of responsibility that closely resembles that of the other two principles. As Henry Shue (2010, p. 111) explains,

> What stands out is that in spite of the different content of these three principles of equity, and in spite of the different kinds of grounds upon which they rest, they all converge on the same practical conclusion: whatever needs to be done by wealthy industrialized states or by poor nonindustrialized states . . . about global warming, the costs should initially be borne by wealthy industrialized states.

The upshot is that wealthy countries like the United States should invest in climate mitigation and adaptation not merely in light of their own narrow interests, but in order to prevent harm to those throughout the world, particularly those in less developed countries who have fewer resources and greater vulnerabilities to climate change. I will proceed under the assumption, then, that any just climate regime must pay particular attention to the plight of the poor: of those least responsible, least able to pay, and most vulnerable to climate harms.

In this regard, I want to consider what conditions might make possible a just climate regime and what conditions might impede it. It is well established that contextual factors influence substantially people's ethical responses. For example, in competitive contexts, such as markets, rational choice models are reasonably good predictors of human behavior; whereas in non-market contexts, they are not, because people tend to behave more altruistically than is predicted by the models (Ostrom, 2010). More generally, *framing effects* influence individuals' responses to moral problems (Kahneman & Tversky, 1984). Experimentally documented framing effects show that seemingly trivial variables such as changes in wording can alter subjects' responses to a moral dilemma (Kahneman & Tversky, 1984). More substantive forms of framing also can have significant effects. For example, framing global climate change in apocalyptic terms through talking of "points of no return," "catastrophe," and "unprecedented changes," may steer the conversation and response to climate change in particular directions (Skrimshire, 2010; Levene, 2010; Amsler, 2010). Levene (2010) worries that talk of "climate crisis" evokes fear and provokes retrenchment and retreat to business-as-usual-style responses, such as the search for technological solutions; nevertheless, the "climate crisis" can be understood differently, as opening up new possibilities for creatively envisioning the future (Levene, 2010; Amsler, 2010).

With respect to climate change generally, and geoengineering specifically, the question I want to consider is this: How do various approaches to addressing the climate problem influence the likelihood of our responding justly over time and of achieving a just, long-term solution? Relatedly, we might ask: What kinds of approaches to climate change will help us see clearly the moral obligations we have—specifically, the obligations to minimize harm and to fairly distribute climate burdens? And conversely, which approaches will obscure our moral view?

SOLIDARITY

Although these questions cannot be fully answered in the course of a single paper, I hope to show that they are highly relevant to the choices we make regarding certain forms of geoengineering as a response to climate challenges. As indicated above, my claim is that large-scale solar radiation management poses threats to moral solidarity, thereby undermining the possibility of a just, long-term solution to climate change. Before turning to this argument, however, I need to say something more about how I understand solidarity and why it is relevant.

Solidarity has as its root the Latin word *solidum*, which refers to the "whole sum" or to the whole. Solidarity describes a certain kind of unity, including unity across difference. There is no consensual definition of solidarity or agreement on what it entails, but William Rehg helpfully describes solidarity "as a kind of group cohesion based on the recognition of a common good (Rehg, 2007, p. 1), and Schwartz (2007, p. 132) characterizes solidarity as "the readiness of individuals to aid other members of a common enterprise." Emile Durkheim, whose theorizing set the stage for many later discussions of solidarity, distinguished *mechanical solidarity*, which is based on resemblance and characterizes relatively homogeneous communities with shared values and beliefs, from *organic solidarity*, a weaker form of solidarity more common in large-scale contemporary societies characterized by a division of labor and a differentiation of roles (May, 1996; Durkheim, 1933). Organic solidarity rests on the interdependence produced by this division of labor, where individuals in diverse roles work together to support and sustain one another in various ways.

Although Durkheim's understanding of solidarity has been helpful, it is not clear that either form of solidarity alone is sufficient to understand the nature and possibilities for international solidarity. The conditions for organic solidarity do seem to obtain in our globalized world: the lines of interdependence are clear, whether in the realm of economic systems, manufacturing, or environmental and natural resource issues ranging from climate change to toxic chemicals to the health of ocean ecosystems. However, the sense of solidarity that people feel with those in other parts of the world seems to be derived, at least in part, from something more akin to the sense of resemblance Durkheim describes as the basis for mechanical solidarity, though the bonds of international solidarity are certainly weaker than envisioned in Durkheim's mechanical model (May, 2007, p. 192). Solidarity across national borders may have multiple sources, but some sense of common interests and shared vulnerability seems to be involved. As Larry May puts it, "[T]he solidarity that many people feel with fellow humans is based on common vulnerability to violence and harm; and the shared interests are based on just these characteristic features of being human as well" (May, 2007, p. 190).

One of the most helpful developments in recent literature on solidarity is a greater focus on its affective dimensions. Jean Harvey (2007), for example, argues that moral solidarity is grounded in an empathetic relationship with others; she is concerned not so much with "*acts* [emphasis added] of moral solidarity" but rather with *relationships* of moral solidarity (p. 23), which depend on empathetic understanding. At the heart of a moral conception of solidarity, for Harvey, is an attitude of concern for others (Harvey, 2007, pp. 23–25). Carol Gould also helpfully develops the connection between empathy and solidarity, arguing that solidarity be understood "as a form of social empathy" (Gould, 2007, p. 149). She writes,

> The solidarity conceptualized here centrally involves an affective element, combined with an effort to understand the specifics of others' concrete situation, and to imaginatively construct for oneself their feelings and needs. (Gould, 2007, p. 156)

Solidarity with individuals at a distance requires a more abstract kind of empathy than is involved in direct interactions with others; it therefore demands greater imagination and perspective taking. Nevertheless, "affective ties of care and concern" remain foundational (Gould, 2007, p. 157). These ties can be strengthened through shared goals and joint projects, as well as through discursive interaction (Gould, 2007, p. 158).

Solidarity refers to a kind of felt connection or unity with others, and this connection may not only *contribute* to moral actions and attitudes but also partly *constitute* them. Thus, we should be concerned about solidarity not only because it can make possible moral choices and just responses to climate change, but also because it helps constitute moral relations among people and nations in the contemporary world. From a normative perspective, we can think of solidarity as an affective moral attitude and a moral disposition: it involves a sense of fellow feeling and attitude of concern that animate a commitment to solving problems for the whole. In the case of climate change, solidarity helps us see that this is a problem we all face, and that we ought to face together, taking into account effects on all people, nations, and ecosystems. This does not imply that the solutions we choose cannot involve diverse initiatives at local, regional, and national levels: breadth of moral concern need not translate into one-size-fits-all solutions. Instead, solidarity involves recognition that climate change needs to be addressed in a way that takes into account the whole, not merely the narrow interests of one group or another. Solidarity commits us to addressing climate change in ways that take seriously those affected by it in diverse locations and situations around the globe and that reflect concern not only for present generations but for future generations as well.

Solidarity bears on both dimensions of climate ethics discussed earlier: a lack of solidarity is likely to contribute to greater total harm and to a more uneven distribution of harm. Without a coordinated approach and broad international commitment to solving the problem of climate change, our "solutions" are likely to fall short: few nations will be willing to bear significant costs without assurance that others will make commensurate efforts. What's more, a fractured, self-interested approach to climate challenges will make more salient inequities in power, leaving weaker and poorer nations to suffer relative to the wealthy and more powerful. Yet lack of solidarity also reflects a deeper problem: it is not simply that a lack of solidarity leads to negative *consequences* in the climate arena; lack of solidarity in a globalized world where our actions deeply affect one another reflects a deficiency in moral relations. Thus solidarity should be understood not merely as a means to the end of fulfilling our obligations to others, but as part and parcel of these obligations. On this view, a just climate regime is one that not only generates a fair distribution of climate-related harms, but one in which those contributing to climate change and those affected by climate change see themselves as part of a common moral community, as standing in moral relations to one another in virtue of their interdependence. I return to these ideas of moral relations and interdependence in the concluding section of the paper. In the next section I explain why geoengineering is likely to undermine solidarity, and thereby, to undermine the prospects for a just response to climate change that averts serious exacerbation of its already-harmful effects.

SOLIDARITY AND SRM

The term *geoengineering* refers to a wide range of strategies that aim to intentionally manipulate the earth's climate. A recent Royal Society report defines geoengineering as "deliberate large-scale manipulation of the planetary environment to counteract anthropogenic climate change" (Royal Society, 2009, p. 1). Under this mantle fall carbon dioxide removal (CDR) techniques, such as tree planting and ocean fertilization, as well as solar radiation management (SRM) techniques, such as increasing the earth's surface albedo, or injecting sulfate aerosols into the stratosphere in order to reflect solar radiation back into space. I focus here on SRM, and particularly on the stratospheric aerosol approach, since this approach currently appears more feasible, effective, and affordable than other radiation management alternatives under consideration (Royal Society, 2009, ch. 3). The general strategy of sulfate aerosol geoengineering is to use small particles to scatter sunlight and reflect it back to space. Hydrogen sulfide and sulfur dioxide are two of the most plausible current candidates for stratospheric injection. Mathematical models suggest that with respect to climate warming, reducing incoming solar radiation by approximately 2% may compensate for a doubling of carbon dioxide in the earth's atmosphere (Royal Society, 2009, p. 29). Evidence for the efficacy of sulfate aerosol SRM comes in part from past volcanic eruptions that introduced significant aerosols into the atmosphere. The sulfate aerosols associated with these eruptions in some instances caused significant global cooling for multiple years, as was the case with the eruption of Mt. Pinatubo in 1991. This eruption introduced approximately 20 million tons of sulfur dioxide into the stratosphere and caused the earth to cool temporarily by approximately half a degree Celsius (U.S. Geological Survey, 1998; Royal Society, 2009).

The dramatic effects of such increases in sulfate aerosol concentrations and the relatively low predicted costs of implementation make sulfate aerosol geoengineering very tempting to some. However, although current models are relatively crude, they predict significant side effects. Most notably, sulfate aerosol injection may change precipitation patterns in ways that exacerbate the effects of climate change on hydrology and agriculture, particularly by altering monsoonal rains in Asia and Africa (Royal Society, 2009; see also Robock, Oman, & Stenchikov, 2008). The addition of stratospheric aerosols also may delay recovery of stratospheric ozone and worsen the ozone hole (Royal Society, 2009). In evaluating the strategy as a whole, we should keep in mind that SRM geoengineering addresses climate change by partially compensating for excess greenhouse gases in the earth's atmosphere and not by addressing directly the root cause of the climate problem. Because SRM does nothing to reduce the total concentration of atmospheric greenhouse gases, sulfate aerosol injection would be a stopgap measure at best, and abrupt cessation of SRM could lead to rapid and dangerous climate rebound effects (Royal Society, 2009; Brovkin et al., 2009).

Insofar as SRM raises the prospects of further climate-related harm as well as concerns over the distribution of harms, whether to engage in sulfate aerosol geoengineering is clearly an ethical question. However, the ethical issues associated with geoengineering are not restricted to those raised by its unintended atmospheric and climatic side effects. There are also concerns over the institutional structures needed to regulate geoengineering research and implementation, and surrounding the fundamental question of who decides when, how, and where to geoengineer, or "who gets to set the global thermostat." At an even more basic level, we can ask whether *anyone* should be controlling the "global thermostat" or whether there are some parts of nature with which we ought not interfere. Concerns about geoengineering as excessive control of the natural world tie closely to objections to geoengineering that appeal to hubris and the domination of nature as problematic aspects of the current human condition (Jamieson, 1996).

Although geoengineering has raised an array of ethical concerns, relatively little discussion has focused on the ways in which geoengineering may influence the prospects for international solidarity, and consequently, for the emergence of a just response to climate change across space and time. Nevertheless, there has been some concern regarding the way in which geoengineering may affect individual and institutional decision-making contexts, inducing choices that stray from the ethical ideal of addressing climate change through substantial, near-term cuts in greenhouse gas emissions—which at this point appear to be the only reasonable approach to reining in serious, long-term harm to humans, animals, and ecosystems as a consequence of global warming. Along these lines, many writers have suggested that geoengineering may introduce a "moral hazard" by alleviating concern to mitigate climate change through greenhouse gas reductions (see, for example, Royal Society, 2009). The idea here is that insofar as SRM offers a near-term buffer against some of the worst anticipated effects of climate change, it reduces the incentive to undertake more difficult measures to prevent those effects, such as making our economies less climate intensive, addressing unbridled growth, and significantly reducing consumption by those in the developed world.

The concept of moral hazard is typically used to describe changes in behavior that increase risk exposure in the presence of insurance against that risk. For example, drivers may be less cautious or willing to take greater risks in their driving if they are fully insured. Of course, it may not *always* be wrong to change one's behavior in response to insurance (Hale, 2009). Moral problems seem to arise when insurance induces behavior that is imprudent, for example, by providing a false sense that incompensable harms resulting from risky behavior are in fact compensable, or immoral, by incentivizing behavior that unjustifiably increases likely or actual harm to others. Insofar as geoengineering may induce complacency about climate mitigation, it may both provide a false sense of security (reducing concern by addressing one piece of the climate problem while failing to address others, such as ocean acidification) and provide reduced incentives to change the business-as-usual patterns of consumption, carbon-intensive energy use, and growth-based economic institutions that are the ultimate drivers of anthropogenic climate change.

However, I am interested not so much in the general relationship of geoengineering to the moral hazard problem, but in specific features of sulfate aerosol geoengineering that raise risks to solidarity. In this regard, sulfate aerosol SRM can be distinguished from standard instances of moral hazard because the "insurance" provided by SRM has a heterogeneous structure. That is, the "coverage" offered by SRM will vary from region to region due to expected differences in regional climate effects. Similarly, the incentives for behaving differently with SRM than without it may differ significantly among nations. More specifically, SRM geoengineering as a response to climate change—and as an excuse to postpone changes in economies, institutions, and infrastructures needed to facilitate significant emissions cuts—is likely to benefit most strongly current generations in developed nations, who have the greatest investment in continuing business-as-usual and who are likely to gain the most and suffer the least from a near-term strategy that focuses on geoengineering. Geoengineering therefore may function like a form of insurance in which an elite group writes the policy and externalizes the greater risk to others who are only partially insured: future generations and those in less developed countries.

Yet we don't need the language of moral hazard to explain how geoengineering threatens moral solidarity and the conditions for the establishment of a morally adequate response to climate change. As explained below, there are three central ways in which sulfate aerosol geoengineering threatens the development and sustenance of global solidarity surrounding climate change:

1. By diffusing energy and focus on long-term solutions;
2. By introducing new inequities that further burden the least powerful while insulating the already advantaged; and
3. By exacerbating intergenerational problems.

The first problem is relatively straightforward: we have limited time, energy, and resources to develop adequate responses to climate change. Geoengineering is a complex and risky endeavor; to be ethically defensible, research and implementation of any large-scale geoengineering scheme will require complex discussions and negotiations. Additionally, it will necessitate the design of new institutions and is likely to engender substantial public debate (see Gardiner, 2010, pp. 293–294 for discussion of the need for new global governance regimes to regulate geoengineering). Thus, there is a serious risk that debates over geoengineering will dilute the effort needed to hammer out a bold, comprehensive, global strategy to address the root causes of the climate problem. What's more, geoengineering may displace a solidaristic, long-term strategy not only by appearing to be a shortcut to solving climate challenges, but also due to what Elke Weber has called "single action bias" (Gardiner, 2009b; Weber, 2006). As Weber explains,

> Decision makers are very likely to take one action to reduce a risk that they encounter and worry about, but are much less likely to take additional steps that would provide incremental protection or risk reduction. The single action taken is not necessarily the most effective one, nor is it the same for different decision makers. However, regardless of which single action is taken first, decision makers have a tendency to not take any further action, presumably because the first action suffices in reducing the feeling of worry or vulnerability. (Weber, 2006, p. 115)

In the case of climate, implementation of a geoengineering scheme may therefore sate the desire to take further action, specifically action to mitigate and adapt. This is likely to be particularly the case for those in the developed world who are invested in doing little more, to whom the positive effects of geoengineering may most directly accrue, and who are unlikely in the short-term to experience significant negative side effects associated with the sulfate aerosol approach. As Stephen Gardiner notes, climate change poses strong risks of moral corruption, "to the subversion of our moral discourse to our own ends" (2010, p. 286). This is a particular danger for the affluent, who generally will be least strongly affected by climate change. As Gardiner (2010, p. 287) puts it,

> [E]ach generation of the affluent is susceptible to arguments for inaction (or inappropriate action) that shroud themselves in moral language but are actually weak and self-deceptive. . . . [I]f members of a generation give undue priority

to what happens in their own lifetimes, they will welcome ways to justify overconsumption and give less scrutiny than they ought to arguments that license it.

SRM geoengineering thus offers a tempting short-term response to climate change that may undermine collective political will to decarbonize the world economy. By diffusing attention and introducing new areas of conflict and debate, SRM geoengineering may attenuate the urgency of reducing greenhouse gases now, for the world community as a whole and for future generations. Rather than focusing efforts on global mitigation and adaptation strategies, geoengineering introduces another set of issues requiring discussion and agreement. In theory, successful international agreements on one issue may positively spill over into the development of agreements on others (witness the general optimism for international environmental problem-solving generated by the Montreal Protocol in response to ozone depletion), but in this case, agreement on geoengineering may be no easier to reach than agreement on a general global climate strategy. And for reasons cited above, even if there *were* agreement on geoengineering, this might not produce further impetus for climate action.

The second problem has already been mentioned: geoengineering is likely to introduce new inequities that undermine the basis for trust and a sense of common purpose that are important elements of solidarity. There are three major areas of concern here. First, there is the problem of uneven effects discussed earlier. Disruptions of monsoons in Asia and Africa are clearly regional in nature, and the ozone hole, while in many respects a "global" problem, has its strongest ultraviolet-radiation increasing effects in far southern latitudes. The second concern involves inequities in *control*. Research and development of geoengineering technologies will almost certainly be concentrated in developed countries with the greatest scientific capacity, and these countries are likely to exert strongest control over the development and deployment of SRM strategies. Given the fact that research and development (R&D) in other areas, such as medicine, has frequently exploited rather than benefited those in less developed nations (see Benatar & Singer, 2010; Glickman et al., 2009 for discussion of ethical issues associated with international medical research with human subjects), those who lack control over geoengineering R&D are likely to be justifiably suspicious regarding its outcomes. Governance structures that ensure greater procedural justice in R&D decision-making could partially mitigate this problem, but the challenges associated with their design and implementation underscore the first point above, that geoengineering research, development, and deployment are likely to bleed resources away from greenhouse gas mitigation initiatives. Distrust generated by real or apparent unilateral development of geoengineering schemes may further undermine solidarity and international cooperation, giving those in developed nations reduced incentives to cooperate and those in less developed nations little reason to trust that industrialized countries will act with concern for their interests.

Relatedly, geoengineering may generate inequities through the introduction of new actors and new risks. SRM research as well as related technological developments will generate personal, institutional, and commercial investments in geoengineering approaches to climate change (see Jamieson, 1996; Jamieson, 2010). Scientists will invest their careers in research programs that aim to understand and operationalize geoengineering; governments and other institutions will dedicate resources to geoengineering; and technology firms will invest capital in the development and testing of geoengineering strategies. New actors and institutions with vested interests in the application and success of sulfate aerosol geoengineering may complicate the debate over climate change and the development of successful and just climate strategies, and they may create momentum to implement SRM strategies despite the risks or before just decision-making procedures are established (see discussions of institutional momentum in Jamieson, 1996, p. 333 and Gardiner, 2010, p. 289). What's more, unilateral geoengineering could be undertaken by an individual nation, or even by non-state actors, without the consent of others or before the consequences of SRM were well understood. This could have devastating impacts on international trust and solidarity. Of course, the challenges associated with the introduction of new actors and interests are not unique to geoengineering. Careers, research programs, and commercial investment have been built around alternative energy development, as well, and I would not argue against alternative energy development on this basis. The difference here lies in the relationship of geoengineering, as opposed to alternative energy development, to the overall climate problem. The unilateral (or "rogue") development of solar power has relatively few risks as compared to rogue geoengineering, both with respect to climate and with respect to solidarity. Unilateral investment in less GHG-intensive technologies may actually *contribute* to solidarity by showing that certain actors (state or private) are working toward decarbonization of energy systems and by making others more able, interested, and willing to participate. Unilateral geoengineering, on the other hand, could not only negatively impact certain people and places, but also deeply undermine global trust. Imagine, for example, a situation in which multiple state and/or non-state actors were working simultaneously on distinct, uncoordinated geoengineering schemes. Such a situation would not only endanger climate stability, it would generate distrust, anxiety, and suspicion, greatly diminishing global solidarity, as well as the prospects for a morally adequate response to climate change as a whole.

Discussion of the first two ways in which geoengineering threatens solidarity emphasized solidarity among presently existing persons, among people dispersed in space, rather than time. However, insofar as SRM geoengineering reduces global intragenerational solidarity, it is likely also to have negative effects on future generations, who depend on the coordinated steps that present generations take to mitigate climate change and to prepare for its most serious effects. Moreover, SRM geoengineering may impact future generations by directly reducing solidarity with future persons: SRM may endanger not only *intra*generational but also *inter*generational solidarity. While some forms of geoengineering—such as carbon dioxide removal (CDR) techniques, which would theoretically slow or reverse increases in atmospheric greenhouse gases—may not damage intergenerational solidarity, sulfate aerosol SRM is likely to do so. Stephen Gardiner (2011a) has emphasized the ways in which climate change invites "intergenerational buck passing" due to the misalignment between the preferences of the present generation and those of each future generation. If SRM facilitates the postponement of climate mitigation, it may turn out to be an intergenerational buck passing strategy *par excellence*. In effect, sulfate aerosol geoengineering offers a tool for temporarily countering some of the most salient effects of global warming—taking the heat off, so to speak. However, it is a strategy that focuses primarily on the present and on immediate needs. If SRM leads us to feel our own vulnerability less acutely, we may fail to take seriously the threats climate change poses to the generations that follow. SRM geoengineering thus may put us at greater risk of a kind of moral corruption that attends to the present at the expense of the future (Gardiner, 2006b). Already, discussions of SRM geoengineering have engendered talk of the "global thermostat" and questions about who gets to set it. While such questions are clearly pressing in the development of any geoengineering scheme, what many have failed to notice is how the thermostat metaphor itself encourages us to think of the climate as something that can be easily modulated with the turn of a dial. And thinking this way deemphasizes important facts about SRM geoengineering: that SRM does not address the steady accumulation of greenhouse gases in the atmosphere, that it poses the risk of extremely abrupt climate change if sulfate injections are terminated, and that unless partnered with serious cuts in greenhouse gases, SRM may lock each generation into an addictive cycle that requires increasingly aggressive sulfate injections to sustain a livable climate. Sulfate aerosol geoengineering thus may lead us to think that each generation can solve the climate problem for itself, simply by adjusting the "thermostat." But such thinking fractures our understanding of the climate problem by treating it in short time slices rather than as a multigenerational challenge. Fracturing the climate problem in this way can in turn fracture solidarity and the foundations for collective action on behalf of present and future generations.

These observations suggest that geoengineering may change the way we think about the climate problem, and in doing so, change the way we conceive our moral relations with others, particularly those in future generations. This insight has special relevance to intergenerational justice, but its implications are more general. If we think of the climate on the thermostat model, it is likely that we will attempt to adjust the climate in ways that provide the greatest "comfort level" to us, here and now. This will affect not only future generations, but those present persons who lack access to the dial, not to mention the plants, animals, and ecosystems throughout the world whose existence depends on a stable climate. Thus the metaphors we use to describe the climate and our efforts to control it may themselves affect solidarity, both directly, by changing the way we conceive of our moral relations with others, and indirectly, by encouraging us to act in ways that further undermine solidarity.

To sum up the key points of this section, geoengineering may influence solidarity in many ways, but three are of central importance. First, geoengineering requires a host of difficult decisions on which parties are unlikely to agree: it raises concerns about the relationship between research and deployment, about governance, and about fundamental ethical questions such as the permissibility of intentional global climate control. Conflicts over these issues threaten solidarity and may distract us from a unified approach to mitigation. Second, the introduction of new inequities—in geoengineering research, through the development of vested interests in geoengineering technologies, and through the direct effects of geoengineering, if implemented—threatens trust and thereby endangers solidarity. Trust facilitates willingness to cooperate and to see the concerns of others as legitimate and important; it is thus an important constituent of solidarity. In the climate arena, trust is critical to establishing an international agreement to control greenhouse gas emissions, both to gain support and participation, and to hammer out an agreement that respects the needs of all parties. Lastly, SRM geoengineering may endanger solidarity by leading us to focus on the present at the expense of the future and by encouraging us to embrace new metaphors that narrow rather than broaden our moral thinking. The "thermostat" metaphor, in particular, may change the way we understand the climate problem and the way we understand our relations with others in regard to it.

When we see climate change as a global problem—multi-faceted, no doubt, but as a single problem facing the world as a whole—we are drawn to address it from this same holistic perspective and to think of ourselves—each of us—as similarly threatened by letting the problem go unchecked. Geoengineering introduces sources of conflict, inequity, and parochialism that threaten to degrade our sense of community with others and lead us toward interest group politics in the climate arena. As we pursue solutions that benefit some significantly more than others, that serve present generations without serving the future, and that are controlled by a narrow elite, we may cease to see climate change as a problem that we can and must face together. We may lose, or fail to develop, a solidaristic orientation that enables us to work collectively to address global warming. Failure to work together can, in turn, further undermine solidarity. Working together can build solidarity—if we approach our collaboration with a shared purpose and sense of our common interests and vulnerabilities. Fracturing influences detract from solidarity by changing our understanding of who we are and how we relate to one another. Loss of solidarity facilitates a shift away from a unified humanity, to an aggregate of nations facing a common threat, to (at the extreme) a loose collection of narrowly self-interested actors with no moral bonds between them.

CONCLUSION

We live in an increasingly interdependent world, in which our actions have significant repercussions that extend widely across time and space. It would be naïve to think that the choices we make—as individuals and as collectives—are morally relevant only to those in our local or national communities, or to those presently alive. We can no longer justifiably act as if the plight of those in other parts of the world or of future persons bears no relation to our own lives and decisions.[3] Though the lines of causation are complex and often indirect—making climate change a difficult and nonparadigmatic moral problem that challenges our conceptual resources (Jamieson, 1992; Gardiner, 2006b)—our choices affect others and put us in moral relations to them. Recognition of these circumstances is, I believe, leading to a gradual transformation in our moral frameworks and a broadening of their scope. Over the course of the last century, increasingly robust international norms have begun to develop—particularly with respect to human rights, but also with respect to the environment. Yet nascent international norms require cultivation: they can be developed and maintained only in a supportive moral climate. We should therefore consider not only the efficacy of various policies and norms within their respective domains, but also their role in building (or undermining) our moral capacity.

In relation to climate change, we need to build our moral capacity through a sense of our common humanity and an empathetic understanding of one another as living beings with interests, needs, and vulnerabilities. Establishing this empathetic understanding and the sense of solidarity associated with it is not easy to do, and some have argued that "global solidarity" is impossible (e.g., Heyd, 2007). Others, however, are more optimistic, and there are encouraging signs. Carol Gould (2007), for example, argues that the mobilization of international support following large-scale natural disasters and growing global justice movements—such as that opposing the trafficking of women—provide examples of transnational solidarity, and she suggests that transnational solidarity with particular others can support the development of a broader sense of global solidarity.

Clearly, more work is needed to understand solidarity on a conceptual level, to clarify through social experimentation and empirical research how solidarity might be fostered, and to develop the connections between solidarity and global justice. But a key first step is to pay attention to how our decisions affect the moral climate. Will our responses to climate change entrench us in comfortable parochialism? Or will they enable us to see ourselves as part of a broader moral community?

A just and ethically defensible response to climate change is possible only if we take coordinated, collective action and overcome the barriers to doing so. Collective action, however, rests on a foundation of trust, reciprocity, and solidarity, and on particular human relationships that embody these qualities. Solidarity across nations and generations will play a key role in developing a morally adequate response to global climate change, and we therefore need to attend both to the direct effects of particular climate strategies and to their role in supporting long-term solutions. Sulfate aerosol geoengineering poses direct risks to global and regional climates, but it also puts the moral climate, and specifically, moral solidarity, at risk. This message is especially pointed for those of us in the United States, as we are citizens of a nation that could exert substantial leadership in the development of substantive international climate accords, though we have historically obstructed progress on climate change. In comparison to Europe, the United States lacks a strong culture of social solidarity, and Joseph Schwartz (2007) has argued that the lack of internal national solidarity carries over into an absence of external international solidarity, leading to isolationism and a realist international political orientation. The path to solidarity—especially for the United States—is clearly uphill. The development of sulfate aerosol geoengineering may make that path steeper still.

Of course, there is no *necessary* connection between sulfate aerosol geoengineering and lack of solidarity, and it is possible to imagine scenarios under which geoengineering could actually *maintain* solidarity, relative to the alternatives. One such scenario would be a case in which the effects of climate change were becoming so severe that failure to geoengineer the abatement of these effects would result in the devolution of the international order, unleashing a global free-for-all, with every country and every generation for itself. I have tried to suggest here that this is *not* our current situation and that there is hope of establishing the solidarity needed to coordinate a global response to the climate crisis. Thus although there may be some circumstances under which geoengineering *would be* the lesser evil (cf. Gardiner, 2010) with respect to solidarity, these circumstances are not yet our own.

There are many considerations to take into account in developing a response to climate change, a challenge which implicates current and past energy systems, global governance regimes, institutional structures at many levels, economic systems, agriculture, forestry practices, and more. In situations as complex as this, it is tempting to focus on technical solutions, to think we can "engineer" our way out of the problem. There is no doubt that we will have to re-engineer our energy systems, many forms of infrastructure, such as transportation and housing, as well as our economic systems, if we are to stabilize the earth's climate. But we need to think not only about how to change *the world* to suit our needs, but how to change *ourselves* so that we can live well together on the earth over time. As we grow increasingly dependent and interconnected with others throughout the world, we are constructing our relations with them through our choices—individual, institutional, economic, political, and technological. If we are not committed to climate solutions grounded in solidarity with others across space and time, our technologies will reflect and reinforce this lack of commitment. On the other hand, if we cultivate solidarity, both through our relations to one another and through our political, institutional, and technological responses to climate change, our choices will not only reflect concern to minimize harm and justly distribute the burdens of climate change today, but establish more firmly an understanding of climate change as a shared burden, a challenge that we must take up together. This, in turn, can lay the foundations for just decisions by generations to come.

NOTES

1. Sincere thanks to Helen Daly, Rick Furtak, David Havlick, Leonard Kahn, Johann Klaassen, Jonathan Lee, Christopher Preston, and the members of the Colorado Springs Philosophy Discussion Group for many helpful comments on an earlier version of this chapter.

2. There is, of course, no *necessary* connection between geoengineering and loss of solidarity; it is certainly conceivable that large-scale SRM could have neutral or positive effects on solidarity. What I aim to show is that conditions are such that neutral or positive effects are unlikely. It should also be kept in mind that even if the effects of large-scale SRM on solidarity were neutral, there are additional risks associated with SRM as well as trade-offs involved in prioritizing SRM as a response to climate change.

3. See Young (2006) for a discussion and defense of a "social connection model" of moral responsibility.

Chapter Two

Might Solar Radiation Management Constitute a Dilemma?

Konrad Ott

In a recent article, Steve Gardiner (2010) has hypothesized that solar radiation management (SRM) might bring future generations into a dilemmatic situation. William Burns (2011) has argued from a legal perspective that a failure of SRM might have grave, and perhaps dilemmatic, implications for future generations. Both articles contribute to the reasons why geoengineering may be a bad idea (Robock, 2008). In this chapter, I wish to make the assumptions underlying the concerns of Gardiner and Burns more explicit. In general, this chapter gives support to such concerns, but it also points at some fragile assumptions in such kinds of argument. Following Gardiner's and Burns's lines of thought, the following argument against sulfate-based solar radiation management (SSRM) begins with the *normative claim* that it is morally either repugnant or wrong to perform a course of action (A) which (with some likelihood) brings future persons as decision-makers into a highly uncomfortable, say, dilemmatic situation (D) on a global scale with great (but different) ills looming on each lemma. Furthermore, I make the *empirical claim* that a course of action (A), which aims at testing and deploying SSRM, entails intrinsic dynamics, serious risks, and a termination problem, all of which might, finally, bring future persons into such a D-type situation as mentioned in the normative claim.

The intrinsic dynamics can be conceived as a *slippery slope*. Therefore, some critical remarks on "slippery slope" arguments are appropriate here. Slippery slope arguments suffer from two open questions: they must hold the two beliefs that a) there really is a slippery slope and b) that the bottom of the slope looks really nasty. Both assumptions are clearly contestable because they are predictive. Slippery slope arguments as such come at the price of two assumptions that are necessarily contested, but these structural features do not render single slippery slope arguments invalid. Slippery slope arguments may be convincing *within* such logical structures. Proponents of slippery slope arguments should be willing to take some initial burden of proof with respect to both beliefs. On the other side, too much should not be required of this initial burden of proof because this might prevent such arguments from "taking-off."

In situations of rising global concern about SSRM, such arguments deserve some attention, if not credit. This chapter takes some initial burden for both the moral and the empirical claims. If one accepts both the moral claim and the empirical claim, it *follows* that the course of action (A) as specified in the empirical claim should not be performed. In other words, it is morally mandatory to omit even the first (real) steps on such an SSRM-route. Most experts would agree that large field tests in the atmosphere count as such real steps. Because there are still better viable alternatives, those should be performed instead.

The overall argument has to be contextualized within the specter of policy options to combat climate change (CCC options). These options are the following:

1. Mitigation: the joint effort of humanity over decades to reduce greenhouse gas (GHG) emissions (saving energy, replacing fossil fuels by renewables, etc.);[1]
2. Adaptation: measures to cope with a changing climate and its impacts;
3. Carbon Dioxide Removal (CDR): measures aimed at removing CO_2 from the atmosphere, aiming for slow but sustainable success and ancillary benefits for nature conservation (afforestation, mire restoration, soil C-enrichment, organic biomass production, combating desertification); and
4. Solar Radiation Management (SRM): measures aimed at manipulating the solar radiation balance of the earth, especially SSRM. SSRM is said to be effective, quick, cheap and risky (see below).[2]

This set of options must be seen as a dynamic political and societal structure. Giving priority to any one of the options within this set has impacts upon the other options. A static *portfolio perspective* might underestimate the dynamic relationship between options. In a genuine portfolio perspective, a portfolio manager decides on a combination of assets in order to maximize the flow of profit stemming from the overall portfolio. All assets are equal insofar as they only count as profit generators. This portfolio perspective usually assumes that assets are discrete and independent from each other. In the highly dynamic triangular affair (consisting of the mitigation, adaptation, and climate engineering [CDR and SRM] options described above) a portfolio perspective simplifies the problem at stake. Roughly speaking, the choice is not between assets but between trajectories. Bearing this in mind, the argument presented here focuses on SSRM as a trajectory rather than as an asset.

NORMATIVE CLAIM

In order to explicate the *normative claim* about an action (A) being morally repugnant, a number of terms need to be specified. First of all, the term "dilemma" is defined here in a broad, weak sense: dilemmas occur if agents have sound (undefeated) moral reasons to do x and sound (undefeated) moral reasons to do y but agents cannot do both x and y. Dilemmatic choices often result in the feeling of moral guilt even if there are not compelling reasons for feeling guilt in such situations. Dilemmatic choices may harm the moral identity of other persons. A dilemmatic choice results in ills, harms, or evils. In our case, both x and y result in comparably bad situations. The ills are of a similar amount but are not identical in quality. Different persons might be affected unevenly, and the number of possible victims cannot be calculated.

The dilemma can be specified as follows: populations may, at some future point in time, have strong reasons to stop SSRM and strong reasons to continue SSRM. However, it is impossible to do both: to continue *and* to stop SSRM in the same period of time.

In order to substantiate the normative claim, six sub-claims (T_n1-T_n6) can be made. The first of these sub-claims (T_n1) asserts that future persons will have some moral rights by which contemporary persons are committed (obliged) to perform or to omit certain courses of action. As Unnerstall has argued (1999), future persons will have rights if there are future persons and if the moral institution of human rights continues. As I have argued elsewhere (Ott, 2004), Parfit's so-called future individual paradox suffers from an ambiguity in the concept of a "particular person." This ambiguity confuses the concepts of an individual and a person. If the confusion is resolved, future persons, whatever their individuality, are to be regarded as bearers of rights. These rights generate some obligations that contemporary persons have the responsibility of meeting. The content of these obligations clearly varies with the overall system of rights that is attributed to future persons, especially with the rights to food, water, environmental conditions, and the like. It is clearly beyond the scope of this chapter to argue for a specific system of rights. It might suffice to claim that any such system should entail (at least) one specific right.

T_n2 then claims that this one specific right is a moral right of autonomous self-determination. This is clearly a liberty right that will be supported by the mainstream of contemporary political liberalism. It is a moral right that has been incorporated into many state and national constitutions in the mode of a legal right.

T_n3 states that this right of autonomous self-determination is either impaired or even directly violated by D-type situations. An action is at least repugnant if it impairs moral rights, and it is wrong if it directly violates such rights. The borderline between impairments and violations is fuzzy and must be determined within constitutional law. Imposing dilemmas on other people should count at least as impairment if not as a violation of the right of autonomous self-determination.

T_n4 entails some suppositions about the relationship between dilemmas and liberty. Autonomous self-determination requires a set of options that are "good" for future persons. Some remaining freedom of choice between "bad" options (such as starving to death in a concentration camp or committing suicide in a concentration camp) is logically compatible with the loss of real self-determination; the remaining freedom to choose one lemma is no "real" freedom. To substantiate this claim, one should distinguish between two different ethical approaches to dilemmas. The first approach is to accept the dilemma as a given, as it has been done in deontologic and analytical meta-ethics. A moral person simply finds oneself being trapped in a dilemmatic situation. The second strategy tries to avoid, if possible, looming dilemmatic situations (for one's self or for other persons) by judging different courses of action. Here, a dilemma is seen not as a given but as an outcome of previous actions. This point is important for the overall argument because it determines the general approach to dilemmas. Imposing dilemmatic situations on others does not establish their liberty; it consumes it. Imposing dilemmas on others deliberately does not merely impose ills; it constitutes evils.

T_n5 claims that discounting future harm is generally permissible if it can be assumed that some items will be less scarce or less problematic in the future than they are today (Ott, 2003; Hampicke, 2003). However, myopia cannot justify discounting future harms. This is true for D-type situations as well. If one imposes a long-term risk upon different persons and some bad event occurs in the future, the bad event will fully count for the persons being affected. Perhaps they may better cope with such bad events if they are better equipped than we are, but this crucial assumption must be likely in order to be valid. This assumption is not very likely in our case. Therefore, it is fair to shift the burden of proof onto the proponents of discounting future D-type situations.

T_n6 asserts that actions that violate the rights of future persons ought to be omitted from consideration *prima facie* (if omitting the action from consideration does not result in an alternative which is too difficult to achieve). The claim supposes that there are alternative trajectories within reach. It is assumed that strong mitigation policies are alternatives that are better or even mandatory and are by no means too difficult to achieve for wealthy countries. The SSRM trajectory is clearly avoidable. I will return to this assumption at the end of this chapter.

EMPIRICAL CLAIM

In order to substantiate the *empirical claim*, several sub-claims (T_e) can be made. These predictive claims refer to the supposed intrinsic dynamics of the A-route (trajectory) toward SSRM and, ultimately, toward dilemma D. Of course, there are many ways to reject such predictive empirical claims. These claims can only be substantiated by some evidence, and people who are critical of them are free to refute them by counterevidence. Most ethicists prefer moral arguments that do not rely on such shaky forecasting and outlooks, but these are unavoidable in our case.

In this section, I split the assumptions into two sets of empirical claims. The first set (T_e1) gives reasons why there will be high pressure in some countries to enter the course of action toward SSRM. The second set (T_e2) gives reasons why a dilemma might loom. The first set of assumptions (T_e1) can be organized in ten different sub-claims:

1. *Tempting Profile*: The route toward SSRM is tempting for all persons rejecting mitigation. The "tempting" profile of SSRM can be characterized as follows: SSRM is scientifically viable, technologically effective, quick to show results, economically efficient in terms of direct costs, and, in principle, unilaterally feasible. It avoids, as Schelling (1996) has argued, all the coordination problems of global mitigation policies.
2. *Rationality*: Elites of advanced industrial societies often act according to a concept of rational behavior that maximizes discounted personal (or collective) utility. Under this approach, long-term risks can be discounted away. This concept of rational choice and the related concept of portfolio management give some support to SSRM. Portfolio perspectives are always about "optimal" choices that maximize discounted utility in a Bayesian risk calculation. Under such approaches, decisions in favor of SSRM might be substantiated.
3. *Big Science*: The tradition of "big science" could be continued and revitalized in an SSRM trajectory. In this tradition (e.g., "Manhattan" project, "Apollo" program), a huge research program might be launched that targets, in the mid-term, large field tests that may show a

real effect on global climate. This research program will likely be well equipped with research money, attracting many scientists. Large field tests will likely be the entry to deployment. To make sense, large field tests must be "low level deployment" (Blackstock & Long, 2010, p. 527). Such field tests will furthermore affect *expectations* of industries and companies that this trajectory will be followed to its completion. Moreover, the SSRM infrastructure and the research groups involved "would lobby to keep the program going" (Robock et al., 2010, p. 531). Once established, big science programs become "too big to fail."

4. *Mindsets*: Such a profile is attractive in a "modern" world, dominated by "hard" science, "big" technology, and "free" capitalist economics. It will be easy to provide SSRM with a nicely designed imaginary framing which fits such mindsets. The imaginary tradition of global climate control can be employed in this respect. Framings, narratives, and mindsets constitute and stabilize influential agency networks.

5. *Moral Hazard*: Expectation that SSRM will be deployed and engagement in mitigation strategies conflict with each other. Whoever expects that SSRM is about to be deployed has a reason not to continue stringent mitigation. Reasons for mitigation strategies gradually fade away. Expectations for serious carbon reduction efforts spiral downwards.

6. *Agency-Network Creation*: SSRM can be seen as common attractor for fossil-fuel industries, the military-industrial complex, consumer democracies, start-up entrepreneurs, patent holders, etc. A strong agency-network is emerging already in the United States which acts actively in favor of SSRM. This network opposes mitigation, adaptation for countries in the global south, and ecological CDR. This network fits well into a political system that has been described as a "managed democracy" by Sheldon Wolin (2010).[3]

7. *Macroeconomic Interests:* In terms of political economy, SSRM is a protective survival strategy for old industries (oil, coal, steel, cars, nuclear, chemical) against new ones ("smart," renewable, low-carbon). Seen as a struggle between competing economies, SSRM is a strategy to protect old-fashioned U.S. economics. Given a global trend toward low-carbon and smart technologies (Jänicke, 2008) and given the industrial and consumption patterns in the United States, SSRM can be seen as a defensive strategy of an economics that lags behind global trends. This strategy is domestically advertised as "global leadership."

8. *Investments*: The further a society progresses along the SSRM trajectory, the more rational it becomes to omit mitigation (and adaptation). As a consequence, there are fewer incentives to invest in renewables and in the production of low-energy commodities. This sub-claim

makes some assumptions about rational investment strategies under an SSRM trajectory. On the one hand, low-carbon energy systems require large upfront investments (WBGU, 2011) that might be discouraging. On the other hand, SSRM outlooks give support to investments in extreme energy projects such as tar sands, deepwater oil drilling, and fracking (Buck, 2011). New economic studies argue that under an SSRM trajectory "green" investments are discouraged and "green lead markets" are weakened (Jänicke, 2012). Technology diffusion is hampered, and energy-intensive lifestyles flourish again. Therefore, the long-term interests of prospering environmental EU industries are not in accordance with SSRM. Different long-term EU energy and climate policy strategies are targeting the emergence of lead markets and the quick diffusion of highly energy-efficient, low-carbon technologies, and renewable energies. There are large investment schemes in solar energy systems to be installed in countries of Northern Africa (e.g., the DESERTEC project). Renewable energies require new "smart grids" throughout the EU. Such "green" investments might be hampered by SSRM prospects. If this is the case, the EU should have a highly critical eye toward SSRM schemes with respect to strategic investment decisions.

9. *Overdetermination*: SSRM will likely be performed because mindsets, interests, agency-network, and rationality-concepts converge toward A. The momentum toward SSRM, once established, is unlikely to be stopped by moral and philosophical concerns. Thus, the window of opportunity for ethical reasoning is at the beginning of this slippery slope, and once it is established, the SRRM trajectory might become self-enforcing. The window of opportunity for ethical reasoning is still open now.

10. *Replacement*: Although the dominant rhetoric at the moment is one of SSRM as a "last resort" or an "augmentation" to the CCC options (under some primacy of mitigation), SSRM might entail an intrinsic dynamic that replaces mitigation and adaptation in the middle and longer run. Mitigation and SSRM stand in opposition not from a logical but from a pragmatic and political point of view. Taking this route may ultimately lead to a choice between giving priority either to SSRM or to mitigation. If SSRM is prioritized, mitigation will not be very substantial and GHG concentrations will remain high on the global scale.

Whereas this first set of assumptions ($T_e 1$) claims that there will be a steep and slippery slope toward SSRM, the second set of assumptions ($T_e 2$) supports the outlook that there might be a genuine dilemma looming at the bottom of this slope. In this second set, I claim that SSRM bears huge risks *and* has a termination problem, which likely implies a D-type situation. The long-term risks include the following: uneven regional distribution of effects, shifting precipitation patterns (decline of roughly 100 mm per year globally[4]), declines in moisture and evaporation, decline of the Amazonian rainforest, weakening of the Indian monsoon, new acid rain, disrupted agricultural production, continued ocean acidification, impacts upon photosynthesis, a delayed recovery of the ozone layer, large and rapid temperature oscillations, and too much (or too little) absorption of sunlight by miscalculation or overestimation of the amount of SSRM.

These well-known risks are surrounded by large uncertainties. For example, the Royal Society claims that "[B]y not reducing CO_2-concentrations, SRM methods would lead to entirely new environmental conditions with impacts on biological systems that are hard to predict" (2009, p. 34). Similar sentiments can be found elsewhere. "It is not clear," write Matthews and Caldeira, "how biological systems may respond to a change in the global relationship between atmospheric CO_2 and temperature" (2007, p. 9952). Political consequences are uncertain as well, especially if SSRM were performed without global consent. Because this chapter does not touch upon the topics of legitimacy and consent, I only claim political uncertainty here.

If any combination of the possible risks materializes, then SSRM might be judged a failure. If we engage in SSRM without sustained mitigation efforts and if SSRM turns out to be a failure, then future generations will be faced with the dilemmatic decision either to further engage in SSRM (=*C*on*tinue*) or to quit the deployment (=*S*top) (D: C v S):

Lemma 1 (C): Continue SSRM despite all the ills it causes.
Lemma 2 (S): Stop SSRM, which, given high GHG concentrations, would imply very rapid climate change with all impacts thereof. (Bala, 2011)

Under the S scenario, successful adaptation becomes extremely unlikely. Because SSRM only treats symptoms and does not cure the causes of climate change, a failure of SSRM will cause increased and rapid climate change. Some middle courses between C and S may remain possible. In case of SSRM termination, "sulfate aerosol density would need to be decreased slowly to avoid ecological shocks" (Robock et al., 2010, p. 531). But such remaining options would be very difficult to control under high GHG concentrations. Perhaps, they only constitute a *tri*lemma. If so, such remaining options would not refute my argument at all.

Analytical ethicists may object that it is not perfectly clear whether D will be a strictly defined dilemma (or trilemma) for future decision makers because it remains possible to assess and compare the ills and evils of the different lemmas in order to determine the lesser evil. If, by definition, one cannot identify the lesser evil in a dilemma, one may argue that it remains uncertain whether the conflict between stopping SSRM and continuing it will be a "real" dilemma. We cannot know today whether future decision makers will be able to identify the lesser evil. Therefore, we cannot know yet whether the situation will be truly dilemmatic for them. Nevertheless, I would hold that a failure of SSRM would result, pragmatically understood, in a dilemmatic situation. Given the magnitude of the SSRM challenge, a purely terminological debate about how to define "dilemma" with respect to future situations is of minor interest to me. It is sufficient to assume situations of hard choices on a global scale.

CONCLUSION

I wish to offer some concluding remarks. The first concluding remark claims that imposing the risk of facing dilemmatic situations on other persons is morally repugnant if there are other viable courses of action available. Imposing the risk of bringing people into such dilemmatic situations is an instance of recklessness. Moral persons should try to avoid imposing dilemmatic situations upon other persons. Such dilemmatic choices do not only affect the material lives of future people for the worse; they also affect their cultural and moral lives. They do not establish real freedom; they reduce liberty. The second concluding remark holds that SSRM does not meet the criterion that failures of technologies should be reversible and should have tolerable negative outcomes even in the worst case (a Rawlsian maximin criterion). Third, a willingness to open up such trajectories indicates how far some persons would go in order to avoid mitigation. This point may be attractive to virtue ethicists.

Finally, there are clearly still better viable alternatives such as "<u>c</u>ontraction and <u>c</u>onvergence plus <u>a</u>daptation" (C&C+A) (Ott, 2010). Mitigation is the only strategy that cures the causes and does not just treat the symptoms of climate change. To make use of a common image employed to critique geoengineering: it is better if a person who is rocking a boat simply sits down rather than trying to find a second person to keep the rocked boat in a shaky equilibrium. This second person is not needed yet.

The components of "C&C+A" are the following:

1. *Strong mitigation* (-50% GHG emissions globally until 2040, with a sharp decline afterwards, a low-carbon world economy in 2100, and a long-term decrease of GHG concentrations), which is possible from a technological and economic perspective (see Stern, 2007; WBGU, 2011). Real opportunity costs of strong mitigation in terms of reduced

growth of global GDP are (very) modest. If GDP is questioned as a crucial measure of human well-being, opportunity costs of postponed and reduced growth lose meaning.
2. *Prudent adaptation* by employing adaptive resilience management in different societal and cultural settings, funding adaptation facilities, using synergies between adaptation and CDR, and generating new models of sustainable livelihoods and development.
3. *Proactive CDR* on a global scale through afforestation, peatland and mire restoration, soil enrichment, biochar, and carbon capture and storage (CCS); creating synergies and win-win-situations between adaptation, CDR, and nature conservation.
4. *Distribution of emission entitlements* on a per-capita-base (with some adjustments), welcoming global redistributive effects and new concepts of development beyond classical official development assistance (ODA).
5. *Enforced technology diffusion* toward India, China, and other rapidly developing nations.

Despite rising emissions, there still is some likelihood of reaching the (politically set) two degrees Celsius target related to global mean temperature, and there might be positive external effects of CDR and adaptation for natural and human systems as well as support for the improvement of livelihoods beyond GDP growth rates. It is beyond the scope of this chapter to assess these auxiliary benefits of CDR, but the topic should be further researched.[5] It is possible to compare a) SSRM with the termination problem (TM) and possible D-type situation and b) C&C+A. All things considered, C&C+A is *far better than* SSRM+TM/D, and "humankind should avoid betting on the fabrication of a silver bullet for shooting climate change" (Schellnhuber, 2011, p. 2). Thus, a moratorium for (large) field tests until 2040 or beyond would be fair to avoid SSRM "hype."

NOTES

1. It seems fair to say that mitigation is the only strategy that cures the causes and does not just treat the symptoms of climate change.
2. For empirical details of sulfate aerosol injections, see Niemeier et al. (2010).
3. Such network-agency-analysis has been done recently in a report written for the "Büro für Technikfolgenabschätzung des deutschen Bundestages," an institution that can be compared to OTA. See Ott et al. (2012 in German).
4. My numbers rely on presentations of Niemeyer and Timmreck of the Hamburg Institute for Meteorology given in fall 2011.
5. See the chapter by Holly Jean Buck, p. 133, in this volume [editor].

Chapter Three

Domination and the Ethics of Solar Radiation Management[1]

Patrick Taylor Smith

> Our scientific power has outrun our spiritual power. We have guided missiles and misguided men.
>
> —Martin Luther King Jr.

Anthropogenic climate change (ACC) falls at the intersection of several difficult moral problems, causing some to describe it as, "the perfect moral storm" (Gardiner, 2011a). It forces us to reconsider our relationship with nature, to grapple with questions of distributive justice, and to confront systematic institutional failure. While all of these moral issues are relevant for any normative analysis of potential policy responses to ACC, this paper concentrates on the ethical problem presented by the obligations we have to future generations. Our actions—both in the creation of and response to ACC—will either threaten or respect the legitimate claims[2] of those who come after us. It is important to take pains to maintain the salience of intergenerational issues when it comes to ACC as the effects of climate change open up the possibility of corruption: we can reap the benefits of carbon emissions knowing that the harms and costs of that behavior will fall primarily on people incapable of calling us to account.

Solar radiation management (SRM), however, offers us a glittering moral possibility: we can meet our obligations to the future and do so at little cost to ourselves. Currently, both advocates and critics of SRM have concentrated on technocratic and instrumental issues: will such a dramatic intervention into complex climate systems "work" without creating detrimental unintended consequences? My paper, however, will be concerned with a further question. Assuming that these technical and pragmatic issues can be resolved, do we still have moral reasons to be wary of SRM? I suggest that we currently have an impoverished view of our moral obligations to future people and that once we acknowledge that we have more robust obligations—in particular, the obligation to refrain from *dominating* future generations—we will see that we have good reasons to avoid deploying SRM even under comparatively ideal circumstances. In other words, we have good reasons to reject SRM that go well beyond concerns about its effectiveness or the possibility of unintended environmental consequences.

CLIMATE CHANGE AND SOLAR RADIATION MANAGEMENT

A strong scientific and public policy consensus has formed around the following set of facts. First, present and past economic and industrial activity has and will, if unchecked, cause the average global temperature of our planet to rise significantly in the coming decades and centuries (IPCC, 2007a; Royal Society, 2009, pp. 1–4). Second, this global increase in temperature will produce ecological effects that negatively affect the life prospects of future people: invasive diseases and species will abound, sea level will rise, and food and water supplies will be endangered (IPCC, 2007c). As a consequence, some societies will be (further) impoverished and some people will suffer from privation, debilitating disease, and death. Third and finally, the best way to respond to this threat would be to engage in a series of reforms of our economic practices and to implement technological innovations with the goal of emitting less carbon, combined with subsidization of (especially poorer) nations that will need to expend resources to adapt to the changes in climate (henceforth, mitigation plus adaptation). Despite this consensus, there is a growing pessimism that the political will exists to implement mitigation plus adaptation (Crutzen, 2006). Recent international conferences have failed to produce any meaningful reforms, and some commentators have argued that there are systematic, game-theoretic pressures that make our current institutions especially resistant to efforts at mitigation (Gardiner, 2003, 2004).

Solar radiation management (Royal Society, 2009, pp. 29–36), on the other hand, apparently offers those of us in the present a way out. Rather than engage in the difficult, time-consuming, and expensive process of either removing greenhouse gases from the atmosphere or reducing our carbon footprint and letting the atmosphere naturally break them down, SRM strategies alter how much solar radiation is ultimately absorbed into the atmosphere (and by the earth's surface) by reflecting a certain amount back into space. That is, SRM strategies increase the reflectivity—or albedo—of our planet, which leads to less solar energy entering or remaining in the atmosphere, which, in turn, leads to a decrease or stabilization in atmospheric temperature. There are a variety of methods for increasing reflectivity, each with varying cost and effectiveness, but the consequence of effective SRM is that we can increase the proportion of greenhouse gases in the atmosphere and yet maintain or decrease global temperature by reflecting more energy back into space. Much attention, for example, has been focused on the possibility of emitting reflective sulfates into the atmosphere (often referred to as sulfate aerosol engineering, or SAG, see Royal Society, 2009, Section 3.3; Gardiner, 2010; Svoboda et al., 2011). It has also been suggested that we place large mirrors at the Lagrange point between the sun and the earth to reflect a certain percentage of the sunlight away from the earth before it even enters the atmosphere or that we reflect more light into space once it enters the atmosphere by using salt water droplets to make clouds brighter (Royal Society, 2009, Section 3.4). Many of these strategies could be instituted at a small *fraction* of the cost of mitigation and adaptation. For example, emitting reflective sulfates into the atmosphere is estimated to cost about 1 percent of that of global mitigation and adaptation.[3] Furthermore, unlike other geoengineering or mitigation strategies, the effects of SRM would be felt almost immediately, additionally obviating the need for costly adaptations even in the near term. Thus, by stabilizing temperature almost immediately, some SRM strategies possibly present an extraordinary opportunity: we can protect future generations from the harmful effects of climate change at a much lower cost to ourselves.

SRM strategies have, despite their promise, considerable—perhaps decisive—problems.[4] Yet, these objections tend to have limited scope due to their reliance on certain common features, which opens the possibility that there are moral problems with SRM that have been left unexplored. First, the most common and powerful objections to SRM tend to be "technocratic" or "instrumental." That is, there are contingent features of both our knowledge and current SRM strategies that undermine its policy rationale: SRM risks detrimental unintended environmental consequences or might altogether fail. While there are also *intrinsic* objections to SRM that argue that there is something inherently arrogant and hubristic about such large scale intervention into the fundamentals of the planetary atmospheric system regardless of their effects, these objections tend to focus exclusively on how SRM evinces a problematic moral relationship between human beings and nature.

Arguments against SRM are not weak simply because they rely on its contingent empirical effects or on how it affects our relationship with nature. But whatever their merits, these objections simply leave open the question of whether there might be other morally problematic features of SRM. It is precisely the thesis of this paper that there are intrinsic objections to SRM that, as opposed to those that rely on our moral obligations to nonhuman nature, rely on our moral obligations to future generations. There are strong reasons to refrain from using SRM that are independent of its environmental consequences or its effects on nature but instead are based on the legitimate claims of future people. Thus, in what follows, I assume that SRM will effectively cool the planet, do so quickly, relieve the most serious consequences of ACC, and will do so without any problematic unintended environmental side effects. Furthermore, I will assume that any problems of international governance have been resolved. The wrongness of SRM will only depend on the kind of act SRM is. Namely, SRM is an arbitrary deployment of power that—especially when compared to the alternatives—dominates the future regardless of the purpose for which that power is used.

THE DOMINATION OF FUTURE GENERATIONS

To begin, we need a substantive understanding of our moral obligations to future generations. Of course, it seems plausible that the trivial desire for larger homes or somewhat less expensive energy is morally problematic if it leads to the deaths and starvation of many thousands in the future, but the nature of the wrong being committed will influence our moral evaluation of the responses to that wrong. An incomplete analysis of the moral foundations of intergenerational justice could lead to a problematic complacency. We could prematurely conclude that SRM is, or would be, permissible because it respected or furthered some values but be blind to the fact that it was failing to respect others. In fact, there are reasons to think that the focus on instrumental or technocratic objections is a consequence of precisely that sort of dynamic. The ethical evaluation of SRM has operated in a philosophical context where the general focus in intergenerational ethics has been on whether, and to what extent, we can harm those who do not yet exist.[5] This univocal focus on harm can warp the more specific discussions concerning SRM in two ways. First, harm does not obviously capture all the important values at play in environmental policy or intergenerational ethics. When it comes to evaluating SRM, questions of distributional fairness or of virtue and vice could very well be relevant. Second, the harm analysis dialectically stacks the deck in favor of SRM, making its justification much easier than it should be. After all, the morally relevant action to take when harming or on the cusp of harming is, generally, to take action to avoid doing so. Once we grant that the real issue is whether and to what extent we harm, instrumental issues are all that remain. What are the most effective and efficient ways to avoid harming future people? And it is here that SRM shines or has the potential to do so, especially given the favorable assumptions operating in this paper. As a result, it is little wonder that current debates on SRM have focused on technocratic issues: if it prevents harm while inflicting little or no harm of its own, what could the objection possibly be? In order to answer that question, we have to explore the possibility that harm—by itself—is an impoverished basis for intergenerational ethics.

As a consequence of this dialectical impoverishment, there have been several recent attempts to characterize our obligations to future generations in more robust terms.[6] While these attempts are promising, my analysis will suggest that domination[7] is a plausible and appealing concept for analyzing our duties to future people and, correspondingly, SRM. As we shall see, domination is not only distinctly suited for the intergenerational context, it also provides us with explicit productive guidance when it comes to SRM. However, first we must be clear about the concept. Domination is often used descriptively. We speak of one poker hand "dominating" another when it has an overwhelming probability of victory. "Strategic dominance" in game theory describes a strategy that is best for the player regardless of the strategies her opponents adopt. Similarly, one can have a "dominating" athletic performance, and Weber described domination simply as the probability that one's commands would be obeyed by someone in a less powerful position. This paper will rely, however, on a *normative* conception of domination, whereby freedom and domination are necessarily at odds. So, to assert that a person is "dominated" by another is to issue a complaint that there are good moral reasons to act to end or change the relationship. To illustrate, consider the following case:

> EMMA AND HARRIET: Harriet is the slave of Emma. Harriet has been the slave of Emma her entire life, living a life where the expectation was servitude. Moreover, Emma is very wealthy and Harriet lives comfortably when not in service. In fact, Harriet's pursuits include activities that would not be materially possible in her ancestral homeland. Over the years, Harriet has come to know Emma's expectations when it comes to cooking and cleaning and does them without command or complaint. Emma, conversely, has great affection for Harriet and generally allows her plenty of personal time to pursue her own projects as long as the chores are done. Of course, *if* Emma were ever to ask her to do something dangerous or inconvenient, Harriet would be forced to do so, or *if* Harriet were to refuse to do her "chores," she would be punished.

There are four things that are important to note concerning this scenario. First, Emma does not obviously interfere with Harriet, and it is certainly not obvious that Harriet's material interests are being ill-served or harmed. Of course, Emma stands ready to interfere with Harriet, but Harriet has shaped her goals and expectations to her subordinate position in such a way as to make such interference unnecessary. In fact, from the standpoint of Harriet's autonomy, she now has choices that she would have lacked had she never been a slave.[8] Second, it is a structural feature of this relationship that Emma does *not* have to interfere with Harriet; growing up under the thumb of

Emma has produced a set of adaptive preferences in Harriet. She has lived under a set of circumstances where freedom is not really a possibility, so her desires have been formed—warped—by that fact to such an extent that she fails to demand it. Third, Harriet is unfree, her autonomy undermined by her problematic relationship with Emma. Fourth, harm-based analyses will generally fail to detect the full wrongness of the relationship between Emma and Harriet, as any interference of Harriet by Emma will be unnecessary when Harriet's preferences have been sufficiently adapted to her subordination.[9]

Emma dominates Harriet. She has superior power and is capable of structuring the choices available to Harriet. Furthermore, she wields that power *arbitrarily*. Her exercise of power over Harriet is dependent on how she happens to feel, what she happens to desire. Even those considerations that work in favor of protecting Harriet's interests—Emma's affection for her long-term house slave in this case—depend on arbitrary facts about Emma. There are no mechanisms external to Emma's own desires that systematically and reliably lead her to consider, take seriously, and respect Harriet's interests when she wields power over her. In this way, domination is a kind of coercive threat writ large. When the highwayman says, "Your money or your life," there is a bare sense in which you "voluntarily" choose to hand over your money. After all, you could have decided to die instead. So, one could say you handed over your wallet "without interference," but this does nothing to undermine the wrongness of the threat. The aggressor has conditioned your pursuit of your life prospects or your rights on his whims. Perhaps the highwayman will treat you honorably, perhaps not, but regardless, you are subject to him. Domination, however, is more structural and more comprehensive than coercive threats, allowing for more indirect constraining of those subject to it. Whether Harriet lives a good life or bad depends, utterly, on how Emma feels and on what Emma chooses, but this is true because Emma occupies a superior social position that brings with it a complex matrix of causal power, behavioral expectations, and privilege. Emma dominates in virtue of being in that relationship and she can command obedience and servility without even intending to. The arbitrariness of Emma's power over Harriet lies in the asymmetry of their structural positions; the institutions, socioeconomic practices, and attitudes that make up the owner/slave relation give Emma the ability to demand compliance without any checks on Harriet's behalf and without any opportunity for Harriet to contest Emma's power. These structural elements further indicate that, unlike a coercive threat that seems to require an intention, one can dominate (and thereby structure the life chances of those in an inferior position) even if one is unaware of it.[10] Possession of the relevant position and the ability to exercise the power implied by that position is sufficient. We could imagine a child king making an exceedingly unreasonable request of a courtier without even understanding why the courtier is so eager to comply. What matters is the *capacity,* the position of arbitrary power, when determining whether one dominates or is dominated. The dominating relationship *remains* problematic as long as the capacity remains. Further, it is unacceptable to have one's life prospects, rights, autonomy, and so on depend on the whims of another powerful person; this can be true *even if* the whims are such that those subject to them are treated well. A "nice" slave owner remains a dominator. A merciful tyrant is no less a tyrant. Of course, it is—in some sense—*better* to live under the heel of someone who likes you and treats you fairly well (you will certainly come to less harm under a benevolent despot than under a brutal one), but it is still the case that you should not live under anyone's heel.

On the other hand, *power*—its exercise, and the corresponding interference in people's choices—are not necessarily morally problematic on this view. Domination is not equated with superior power but rather with the capacity to wield one's superior power without any external or systematic constraints to protect the interests of those subject to it. An officer of a modern constitutional democracy will wield vastly superior power over private citizens, but the appropriate institutional design can make it such that the power must be wielded reliably in the name of the common good and the autonomy of free and equal citizens and with externally effective checks that ensure accountability to those subject to it. The individual member of a local police force in a just state does not wield power as a private individual who may lock people up or punish them in virtue of her paramilitary training and weaponry. Rather, she wields power as a public officer, constrained by the rule of law available to and applying equally to everyone, enforced by mechanisms and accountability that move far beyond the particular whims of the individual police officers: internal affairs investigators, written constitutions, independent judiciaries, criminal and civil proceedings against wayward officers, citizen review boards, free, democratic elections of civilian superiors, and rigorous professionalization. So, while it is true that there are structural, institutional, and social mechanisms that make a police officer more powerful than the average citizen (in fact, given the respect, institutional and political backup, and social position of the police, they have considerably greater power than citizens even with equivalent training and weaponry), this power is correspondingly constrained by forces external to the individual officers. And importantly, these constraints point in one direction: they pressure the police to act for the common good rather than simply their own. Insofar as constitutional liberal democracies succeed in directing power appropriately such that it serves the people subject to it, they are non-dominating over their citizens no matter how much power they wield. This is why domination is an *institutional* failure.[11] It is very hard to see how a person can serve as a check on *herself* or how a despot self-guaranteeing their own accountability is not a kind of ersatz, non-comprehending simulation of genuine answerability. This suggests that there are *two* primary ways to repair a dominating relationship. One could work to equalize power—either by taking it away from the dominator or by giving it to the dominated—or one could make it such that the superior wields power non-arbitrarily via institutional reform or by reforming the relationship more generally.

This analysis is directly applicable to SRM as a response to ACC because SRM exacerbates the current domination of the future by the present. While the next section will argue that SRM is inherently more dominating than mitigation plus adaptation under even ideal circumstances, the remainder of this section will examine how SRM will exacerbate, deepen, and recreate the *current* domination of the future by the present. To begin, human beings—if we research and develop SRM capacities—will be able to intervene into the fundamental processes of planetary ecology to an extent unfathomable until very recently. Without institutional reforms that ensure that the power of the present is deployed non-arbitrarily, any increase in power correspondingly deepens the relationship of domination. It is irrelevant that this new power could be used to benefit those subject to it, and it would be equally irrelevant even if the negative side effects of its deployment were relatively *de minimus*; dominating power is not immune from moral criticism simply because it might be used well or because it might not be all that harmful. Furthermore, as we shall see, SRM deepens the nature of the domination of the future by the present by undermining what few checks remain on the present's power.

However, one might think that there is something quite odd about the idea that SRM increases the arbitrary power of the present. It is, of course, true that we wield considerable power over the future when we engage in climate change or in SRM. Moreover, it would seem that we wield maximal and unchecked power over the future, and, unlike most social and political relationships, we *necessarily* do so. After all, even hunter-gatherers could always have chosen to burn their crops or stop having children and there would be nothing that future generations could do; SRM adds essentially nothing to the underlying power dynamic. If what matters is the capacity to wield power arbitrarily, then we have always had and always will have that over the future. Thus, why should we think about the *particular* wrongness of climate change or SRM in terms of domination? Call this the *inescapability* objection. If one cannot avoid a wrongdoing, then it makes little sense for that wrongdoing to play a large role in our practical deliberations. If we dominate the future *regardless* of what we do, then what does it matter that we do so with ACC and SRM? The answer is to note the difference between *formal* and *substantive* power and to see that the latter poses a much greater dominating threat. Our formal power, derived from the causally asymmetric nature of our relationship with the future, may well be ineliminable.[12] But while our formal power has remained the same (the basic causal nature of the universe is unchanged: the present influences the future but not vice versa), our substantive power has grown and grown in a particularly dominating direction. What this means is that the *scope, magnitude,* and *nature* of our power has widened, deepened, and changed. While we have always been able to influence the future, early humans could usually only do so in comparatively small ways over comparatively small regions.[13] In other words, human causal influence was limited, in effect, in space and in time. Technological advances and industrialization have quite radically changed this. Now, even comparatively small groups of humans can deeply impact the ability of an ecosystem or even the entire planet to support human life and can do so fairly quickly. While formal power has been constant, we have become more substantively powerful and as we have done so, the dominating nature of the relationship has deepened and the normative complaint domination represents has become more urgent.

More importantly, the fundamental *nature* of that substantive power has changed in a much more dominating direction because technological and economic changes have undermined one of the few genuine checks on the present's treatment of the future. Before industrialization, negatively influencing the future generally—though perhaps not always—required the harming of the interests of the present. In other words, there was usually a comparatively close alignment between actions that served the interests of the present and actions that would sustain the future. Hunter-gatherers who ceased having children would suffer greatly in their old age. Members of an agricultural society who burned their crops or ceased to maintain their irrigation infrastructure would soon find themselves starving. Even today, many of our own most dramatic examples of substantive power—nuclear weapons and the corresponding ability to make the planet instantly uninhabitable—have this inherent check: we can destroy the future only by destroying ourselves. In other words, for much of human history, the present's substantive power over the future was usually weakened by a strong internal check based on the current generation's interest. This check, for much of human history, provided an external constraint on how far present people could exploit their position to benefit themselves to the detriment of the future. While this check was too problematically contingent on economic practices and technology, never sufficiently reliable, and based on an insufficiently tight linkage between future and present interests to *eliminate* the arbitrariness of the present's power, it certainly did *limit* it.[14]

Industrialization undermined this dynamic, redistributing the benefits and burdens of the present's exercise of substantive power and removing this internal check. In other words, industrialization creates *front-loaded* goods with *back-loaded* costs (Gardiner, 2003). When we emit carbon, we gain the benefits of increased economic growth and cheap energy, but the negative externalities of that resource consumption are temporally extended across the generations with the bulk of the costs coming well after we die. Moreover, this logic is recapitulated with each generation, as the costs of the *current* emissions will not be felt until later no matter how badly the present suffers from the emissions of prior generations. Thus, the present generation's deployment of substantive power instantiated by the large-scale emission of carbon lacks the internal check characterized by the substantive power of pre-industrial generations. This dynamic explains the radically different reactions to the otherwise similar environmental problems of climate change and ozone depletion. When we discovered that the use of chlorofluorocarbons (CFCs) was depleting the earth's ozone layer, we were faced with an environmental calamity where the present generation would *suffer the ill effects* of their behavior. A depleted ozone layer would lead to higher cancer and cataract rates for the presently existing people who were benefiting from the CFC use. As a consequence, the present generation acted to reduce CFC use dramatically and repair the ozone hole, which has now stabilized. The use of CFCs, which would make the lives of future people worse, exhibited this kind of internal check. Climate change, however, does not, or does so to a much lesser extent. The benefits of carbon emission are high and the significant costs will not come for some time, especially for those who emit the most and have the resources to easily adapt. Just as importantly, for each generation, the costs of carbon emission are back-loaded: the effects of the previous generation's emissions will cause negative consequences to the present, but the present does not harm itself by its own emissions.

To tie these threads together, we increasingly dominate future generations. We are in a position of vastly superior power, not just formally but substantively. We can and do exercise that power to condition the legitimate claims of future generations on the satisfaction of our particular desires. And we do so increasingly arbitrarily, as even the imperfectly effective substantive checks that have constrained the present have been worn away technologically; we currently lack the relevant institutional checks designed to force us to consider the interests of future people. This analysis thus provides a response to the *inescapability objection*. Formal power over the future is inescapable, but substantive power (and the corresponding level of domination) can rise and fall with technological development and the presence of internal and external checks on that power. Some technologies encourage and deepen intergenerational domination; others lessen it or are neutral. Finally, the power of the domination complaint seems actually to *increase* as the relationship between dominator and dominated becomes attenuated because increased interaction strongly tracks and makes possible all sorts of negotiations between the dominator and dominated. As the opportunities to check the power of the dominated decrease, as the demands for answerability and accountability become weaker, domination deepens. Whereas the causal influence operates in only one direction, this is no obstacle to analyzing our relationship with the future in terms of domination and we can understand different levels of domination in terms of substantive power. The development of SRM represents an increase in substantive power and, thereby, an increase of domination on several fronts. SRM technologies would allow us to intervene in the global environment and alter fundamental atmospheric processes in a way that we could not previously. Thus, the magnitude and scope of the present generation's causal capabilities would increase upon the development and greater sophistication of SRM techniques, just as industrially developed generations had greater capacity to influence the future when compared to those who came before.

More importantly, SRM intensifies and recapitulates the cost-benefit dynamic of industrialized ACC. As we saw, one important traditional limitation on the substantive power of the present has been an internal check based on self-interest. ACC is an especially difficult political problem because that check disappears: we gain the benefits of cheap energy by imposing costs on those who do not yet exist. In other words, we become better off economically than we otherwise would have been had we not emitted carbon. SRM intensifies this structure by making it *even cheaper* to take the benefits of carbon emission and to impose costs on the future, but it does so along a moral baseline: there are costs imposed on the present in virtue of the *moral* responsibility we have to respect the legitimate claims of posterity. The cost of mitigation plus adaptation is—in a sense—the moral price of our profligate emissions behavior, and the perceived obligation to pay that price can itself serve as a check on our decision making or on the behavior of similarly situated people in the future (just as the knowledge that one be wracked by guilt or would need to make a humiliating public apology to make good some wrongdoing can motivate one not to do wrong in the first place). SRM undermines this last deterrent by creating a perception that one's moral obligations to the future have been satisfied, creating an opportunity to escape the moral as well as the economic costs of our emissions. In any case, SRM intensifies the problematically dominating nature of ACC simply by making it easier and less costly for us to influence the future. Just as industrialization made it possible for the present to avoid the *economic* costs of injuring the future, SRM makes it possible for the present to cheaply avoid any of the perceived *moral* costs for recklessly endangering the legitimate claims of the future.

In addition, SRM itself exemplifies the back-loaded cost dynamic. While it is true that SRM can be easily "un-deployed" in the sense that we can stop putting sulfates in the atmosphere or blow up our space mirrors and the albedo of the planet will return to "normal," SRM as a response to climate change creates a significantly path-dependent dynamic. That is, if SRM is not accompanied with other policies to decrease our greenhouse gas emissions, then any reversal of a SRM policy in the future will lead to quite rapid climate change as the decreased albedo interacts with an even greater amount of greenhouse gas (Ross & Matthews, 2009). Thus, the decision to engage in SRM now will present distant future people with a difficult choice: they must accept the need to continue with SRM forever, or they will be faced with the need to engage in far more expensive mitigation and adaptation efforts than demanded of the present *before* they can cease SRM. SRM imposes greater and escalating costs on the future. Furthermore, it seems likely that the costs of mitigation plus adaptation will fall most heavily on the present and those directly following it, so this means that SRM will benefit the present and the near future by imposing progressively greater costs on the far future. Of course, one could argue that these termination costs are only relevant if there is some expectation that SRM will cease to be effective or that future generations will fail to pay the relatively trivial costs of upkeep. So, the next section will argue that SRM remains dominating—especially when compared to mitigation plus adaptation—even if we set aside these issues concerning termination costs.

SOLAR RADIATION MANAGEMENT AND INTRINSIC DOMINATION

SRM would—in our current institutional context—be a particularly egregious form of domination. Yet, the dominating features of SRM described above only relate to increasing the extent, scope, and economy of the present's substantive power. But increased power is only dominating when the social and institutional context is such that it is wielded arbitrarily. There is, then, the further question of whether SRM would be dominating even if we reformed the *intergenerational* institutional structure such that the present could usually deploy power non-arbitrarily over the future. So, let us suppose that we have engaged in the relevant institutional reforms[15] that require the current generation systematically and reliably to take the interests of future generations into account in their deliberations. Would there still be a problem with regard to SRM? In other words, is it the case that SRM is inherently dominating,[16] such that its deployment must be arbitrary? The rest of this paper is dedicated to demonstrating that the answer to these questions is "yes."

The first step is to realize the moral relevance of SRM as a response to a *human-made* threat. Consider Christopher Preston's (2011) intriguing thought exercise:

> If it were the case that humans had not released large quantities of greenhouse gases into the atmosphere and that the world was warming naturally at the same dangerous rate that it is now warming due to anthropogenic greenhouse gases, would the environmentalist presumption against geoengineering be enough to preclude taking measures to prevent the impending change? (p. 471)

The acceptability of SRM does appear to depend crucially on whether the atmospheric warming to which it is a response is natural or man-made. But this is puzzling in the light of the typical objections to SRM. We may still worry about the unintended consequences of SRM or the hubristic tinkering with nature regardless of whether it is responding to natural or artificial phenomena. The key to resolving the puzzle is to see ACC as not merely human-made but as a human-made threat. Present and past generations have taken advantage of their dominating position to impose harmful consequences on those weaker in order to benefit themselves. The fact that SRM is a response to a human-made threat needs to be taken into account when evaluating it. Let us illustrate this with an example. Suppose a person comes across a hiker injured in an animal attack and there are two ways to get her to safety: one path that will be better for the injured but more dangerous and painful for the rescuer and another where the dangers are reversed. In saving the hiker, it seems that technocratic questions take prominence. What path instrumentally serves the interests of the two people? One method of rescue might be better for the rescued but so much worse for the rescuer that she might legitimately reject it. However, if the hiker had been injured because the would-be rescuer chased her off of a cliff, the nature of the rescuer's obligations to the injured would be quite different. It matters in determining what the rescuer is morally obligated to do that, in the latter case, the rescuer has used her power to cause or threaten the harm that leads to the need for the rescue in the first place.

But how is this relevant to the evaluation of SRM? What could be the problem if, as this paper has assumed, there are no serious negative environmental side effects of SRM? Why not just continue with SRM forever? If SRM works perfectly, why not simply choose the cheaper, easier option? The answer lies in the moral difference—in terms of domination—between the *blocking* of a threat one has created and the *elimination* of that threat. In the context of ACC, SRM represents the former and mitigation plus adaptation represents the latter. The reason has to do with the termination problem discussed at the end of the last section. Mitigation involves the gradual *removal* of the greenhouse gases that cause the damaging increase in global temperatures. As time goes on, the threat of the warming decreases until, eventually, it can only be recreated with difficulty as greenhouse gases in the atmosphere return, or come close to, pre-industrial levels. Furthermore, adaptations strategies have the effect of making future people less vulnerable to the threat of climate change. Mitigation and adaptation have the overall goal of removing the threat of ACC related environmental and economic damage. The longer mitigation plus adaptation remains in effect, the less vulnerable future people are to climate change, and the more difficult it will be for any future generations to restart the morally problematic dynamic that has led to the looming catastrophe of climate change. SRM, by contrast, does not eliminate greenhouse gases but rather allows them to continue building as it prevents those gases from creating their detrimental effects. In fact, if one were to cease engaging in SRM, the threat of ACC would be much *worse* as the warming would be vastly accelerated. SRM thus *blocks* the threat of ACC, but the dangers of climate change do not go away. This distinction is the fundamental feature that makes SRM an inherently dominating policy option, especially when compared to mitigation and adaptation.

Generally speaking, we might be worried about the mere blocking of a threat because of the unreliability and expense of the blocking mechanism. If there were a massive run-off caused by topsoil depletion, we would probably prefer a more sustainable agriculture as a response when compared to a dam or a levee because we would worry that the dam might give way or that failure to fix the underlying problem will lead to ever more expensive upkeep. Reforming agricultural practices might be initially more expensive, but over the long run it would be a much more attractive strategy. Similar worries can and should play a role in our evaluation of SRM. As the amount of carbon in the atmosphere increases unabated, more sulfate will need to be placed in the upper atmosphere or more mirrors added or adjusted in order to reflect more sunlight. This, of course, increases the possibility of catastrophic failure and harmful side effects. It also leads to increased expense so that future generations will become increasingly burdened by the demands of the continuing SRM.

However, *even if* we assume that there will not be a catastrophic failure or problematic side effects, blocking an intergenerational threat is substantially worse—in terms of domination—than eliminating it. There are, fundamentally, two reasons for this. First, both blocking and eliminating a contemporaneous threat are consistent with the threatening agent being able to re-instantiate that threat at a later date: the threat can be re-made or the block removed. This is why, in intragenerational cases, it is insufficient simply to block or eliminate a threat: we must create institutional safeguards to structurally guarantee that the threat will not be re-made at the whims of the threatening agent. In the intergenerational case, however, eliminated threats cannot be simply re-instantiated by the same agent. Once the current generation eliminates the threat of carbon emissions, that generation's threatening influence over the future via ACC will come to an end. However, if the current generation merely *blocks* the threat of ACC, the present will continue to threaten the future. Later generations may take up that threat by refusing to maintain the SRM infrastructure or by continuing with similarly reckless environmental behavior that—at the moment—has been insured or covered up by SRM. This possibility is especially salient in cases—like sulfate aerosols—where the costs of maintenance will go up over time if carbon emissions are not cut.

However, this seems to ignore the temporal nature of climate change. It is, of course, true that once *this* generation begins eliminating the threat of carbon emissions and dies out, *it* will not be able to threaten future generations. But equally obviously, the *next* generation—or the one after that—will be able to re-instantiate the threat by rescinding all the reforms that the prior generations enacted. This appears to be analogous to some future generation deciding that the cost of SRM is too great and choosing to stop. In fact, one might think that the lesser costs of SRM would decrease the likelihood that future generations will re-engage in risky emissions behavior; its maintenance is less expensive. These appearances are deceiving because mitigation plus adaptation is not *simply* a strategy for eliminating carbon from the atmosphere as opposed to, say, carbon dioxide removal techniques (Royal Society, 2009, pp. 9–21). Rather, mitigation plus adaptation is a complex suite of large-scale reforms to social, political, and economic systems. Agricultural patterns, transportation infrastructure, industrial production, and energy consumption will all need to be extensively changed in order to produce an economy with a sustainable level of carbon emissions, given the population of earth and our expectations about what constitutes a decent life. These changes will create new expectations, new interests, and new patterns that will solidify and form the basis of the life plans of those born into them. They will create a new kind of path-dependency, a new systematic pressure in favor of producing less carbon instead of more. And once these patterns come to be, they will be *difficult* to change as they will have incorporated large-scale capital investment and sunk costs and will form the foundation of economic planning and expectations. In other words, the large-scale reform of our economic system demanded by mitigation plus adaptation will have the effect of structurally re-inscribing the pre-industrial check on a particular sort of risky environmental behavior: the emission of greenhouse gas. And this is not a contingent feature of mitigation plus adaptation: it is precisely how the response is supposed to work. Mitigation plus adaptation is not merely an elimination strategy; it is an elimination strategy that makes it more difficult to re-instantiate the threat in the future. From the standpoint of domination, a response to a threat that not only eliminates it but also creates a structural check against that threat has to be preferred over a response that merely blocks the threat, leaving it in place to be used or exploited later.

Thus, mitigation plus adaptation, when considered in light of the value of non-domination, must be considered intrinsically superior to SRM. We already had strong moral reasons to mitigate rather than deploy SRM, but these reasons are *strengthened* by the particular dynamic between SRM and ACC. We have good reason to reject the deployment of SRM as an answer to climate change even if we grant that SRM will successfully resolve all the harmful environmental consequences of ACC.

Yet, one might think at this point that this argument only applies to the idea that SRM will be the primary or only response to ACC. The natural question to ask is: How does this argument apply to an SRM deployment that is merely one component of a complex cocktail of responses that include carbon removal techniques, adaptation, and mitigation? On this view, SRM would "buy time" to allow other strategies to take effect. Whereas we might easily understand why eliminating a threat is superior to merely blocking it, it is harder to see the moral problem with blocking a threat *while* working to eliminate it. Let us call these "hybrid" responses to ACC. So, while the argument above may have shown that any deployment of SRM must be accompanied by eliminative strategies, it has not shown that deployments of SRM in hybrid responses are unacceptable.

There is certainly something attractive about this line of reasoning. There is a kind of division of labor where SRM stops or slows the contemporaneous human and ecological effects of ACC while other mechanisms deal with the long-term effects with the goal, perhaps, of making SRM eventually unnecessary. However, there are several caveats and reasons for skepticism. First, not all eliminative strategies are alike. The reason that mitigation plus adaptation is so strongly non-dominating is that it eliminates the threat of ACC while also creating a structural check on that threat's reproduction. An ACC response cocktail that included only SRM and carbon removal techniques would probably fail the domination test when compared to mitigation plus adaptation because the sequestering or scrubbing of carbon does not provide any structural check on the ability of future generations to recreate the threat of ACC should they decide that such sequestering or scrubbing is too expensive or difficult. So, the intergenerational domination analysis can, at the very least, delineate which elimination strategies must be included in any ACC response suite. Second, the rhetoric that accompanies hybrid strategies is problematic, relying on the need to accept the "lesser evil" or "buy time" in an "emergency." It is important to interrogate these arguments, asking whether we truly need to buy time or whether we really are in such an emergency (Gardiner, 2010). Third, there is a worry that by engaging in SRM we might undermine the motivation to engage in mitigation plus adaptation. Once we have blocked the threat, we have removed much of the incentive to eliminate it. This is especially so because SRM will have the effect of doing much more than other, slower strategies to block whatever effects of ACC we may feel in the present. The movement to deploy SRM is predicated on the general worry that the present will lack the political motivation to instantiate difficult and expensive eliminative strategies. It is hard to see how decreasing the urgency of the threat of global warming will result in a greater political will to tackle it. Thus, the deployment of SRM is more problematic insofar as it undermines the likelihood of the deployment of non-dominating eliminative strategies. The permissibility of SRM would require three things: it would have to be part of a suite of responses that included non-dominating eliminative strategies, it would have to be genuinely necessary for achieving important policy ends, and it would have to have no effect on the political will to instantiate non-dominating elimination strategies. This is a high standard, though perhaps not completely insuperable.

CONCLUSION

My conclusion is both comparatively robust and somewhat narrow. SRM as the sole or primary response to climate change is deeply problematic because it instantiates, deepens, and perpetuates our domination over the future. SRM dominates in a way that mitigation plus adaptation does not; that domination is an inherent feature of how SRM operates as a response to climate change. This does *not* mean that SRM can never be morally permissible in the context of climate change. As part of a general climate change response-cocktail, a limited use of SRM *might* make it easier to mitigate or adapt and could be *temporarily* justified on those grounds, assuming we could be reasonably confident that its temporary use as a part of the solution would not evolve into something more permanent and all-encompassing. On the opposite end of the spectrum, the general wrongness of SRM might be overcome if it were absolutely necessary to prevent a disaster. For example, if we had some reason to believe that warmer temperatures would *very soon* lead to a massive and massively destructive climate disaster, then avoiding the catastrophe of human extinction would justify the deployment of SRM. However, even in these cases we must remain responsive to the threat that SRM poses, both to the life chances of future generations and the moral corruption that such easy domination engenders in those who wield power.

NOTES

1. This paper was written and made possible by a National Science Foundation (NSF) summer research grant at Northeastern University (NSF Grant No. SES # 0609078). I would like to thank Christopher Preston, Ron Sandler, Steve Gardiner, and my summer fellows for reading various versions of this paper and providing invaluable assistance.
2. The locution "legitimate claim" is used to sidestep skepticism about intergenerational justice *tout court*. This paper will simply assume that we do owe future people a reasonably sufficient physical and environmental infrastructure necessary for decent lives.

3. Estimates of mitigation plus adaption usually run around 1–2% of global GDP, or $700 billion. Sulfate aerosol costs are usually estimated as being somewhere between two and ten billion dollars (Barrett, 2008).

4. First, we do not know with any certainty what the unintended side effects of albedo manipulation will be, and they will likely have regionally-variable effects (Svoboda et al., 2011; Corner & Pidgeon, 2010; Gardiner, 2010). Second, the effectiveness of these strategies has, for the moment, been estimated on the basis of established climate models, but it is difficult to see how they can be genuinely tested short of deployment (Robock et al., 2010). Third, SRM strategies are only able to block the harmful consequences of global *warming*, but they do nothing to prevent the negative effects caused solely by increased carbon in the atmosphere, such as ocean acidification. Fourth, while SRM strategies are cheap enough that a single nation or even a corporation could use them, the global consequences of atmospheric sulfate emission or space mirror deployment would seem to require globally-oriented political institutions that are, at best, inchoate or, at worst, non-existent. It seems naïve to think that the nations of the world would idly accept such a radical and unilateral intervention into the climate (Corner & Pidgeon, 2010). Finally, some have argued that SRM represents a kind hubris, an arrogant desire to dominate nature.

5. For example, Derek Parfit (1984) presents his now famous "non-identity problem" in terms of a puzzle about how our commonsense notions of harm can apply to cases where we are deciding who to bring into existence. Respondents to Parfit (e.g., Woodward, 1986; Caney, 2008) have equally focused on when and how we harm those we create, or they have suggested new, harm-like concepts that undergird our commonsense intuitions that we do harm when we create people under bad circumstances.

6. Some have suggested that we exploit future generations with our current emissions (Bertram, 2009; Rendall, 2011), some have attempted to extend notions of community to cover the distant future (Thompson, 2009), while others have suggested that we can extend the reasoning and logic of the social contract to apply intragenerational concepts of justice to future people (Gauthier, 1986; Heyd, 2009; Kumar, 2009). I tend to think that domination is theoretically superior, but that is beyond the scope of this paper.

7. My discussion here owes much to, and is meant to be agnostic between, the prevalent views of domination in the literature. The major views are represented by Pettit (1997), Lovett (2010), and Young (2002). My first inspiration to apply domination to the issue of intergenerational ethics comes from Nolt (2011a).

8. Perhaps one could argue that Emma could do even better by Harriet's interests by giving her the same opportunities and not enslaving her. But it seems unlikely that Emma would generally be obligated to bring Harriet from her homeland to serve her interests (unless the homeland was sufficiently impoverished or violent). So, we are still left with the question of what is problematic with Emma's conditioning of Harriet's enjoyment of superior material prosperity on enslavement.

9. The harm theorist could say that Harriet has been *harmed* by living under oppressive conditions and having her preferences so shaped. This is just another way of saying domination is a harm, but then we still need an analysis of domination in order to determine when that particular kind of harm has occurred.

10. Does this mean that, for example, we dominate when we drive our cars because we threaten pedestrians? The answer to this question will be determined by the social context. Where the rule of law operates and there are robust and effective legal protections in place that will track the interests of pedestrians and check drivers, then driving would not be dominating. If those protections did not exist, driving might then be an instance of domination.

11. The extent to which institutions themselves depend on a virtuous citizenry is, of course, a large question in the history of political thought. All I wish to suggest here is that individual virtue is never sufficient to prevent domination, but it may very well be necessary.

12. This does not necessarily mean that intergenerational *domination* is ineliminable as it remains possible that we could wield that formal power non-dominatingly.

13. This was usually but not always the case. The extinction of the North American megafauna was a technologically driven event that considerably influenced the future, though even this extinction happened at a rate considerably slower than the major extinctions or near

extinctions brought on by industrial age capabilities. So, the individuals and their descendants who crossed the Bering Strait did dominate the future, but it is possible that they did so *blamelessly* because they could not be fully aware of the consequences of their actions.

14. The debate between Neo-Republicans and Neo-Kantians (Hodgson, 2010, Section V) about whether checks based on public, political institutions are to be inherently preferred over those predicated on market and non-public mechanisms is beyond the scope of this paper, but Neo-Kantians would find the internal, economic check especially problematic.

15. For example, we could create legal mechanisms that are selected to represent the future similar to the *guardian ad litems* designed to protect the interests of children or legislative offices similar to the tribunes of ancient Rome. There is much more to be said on this topic, but it is beyond the scope of this paper.

16. An example of a dominating action done by generally non-dominating institutions would be a constitutional, democratic state restricting the ability of gay couples to marry or adopt. That kind of unjustified, unequal treatment can only be done arbitrarily.

Marginalized, Vulnerable, and Voiceless Populations

Chapter Four

Indigenous Peoples, Solar Radiation Management, and Consent

Kyle Powys Whyte

Funding research on solar radiation management (SRM) is now a policy option for responding to climate change due to the perception that international abatement efforts are creeping along too slowly. SRM research presents a range of problems concerning consent for Indigenous peoples.[1] Indigenous peoples' landscapes may risk rapid, unforeseen changes that will force communities either to respond under great hardship or migrate elsewhere. Since the science and engineering behind SRM are esoteric to nonexperts, legitimate concerns arise about transparency and procedural justice. Indigenous peoples may also contest the very idea of human "control" of global temperatures. In this paper, I will examine what it would take for parties interested in funding, designing, and carrying out early SRM research to fairly respect members and leaders of Indigenous peoples in their current discourses.

Ethical concern is warranted. Indigenous peoples have yet to be addressed responsibly about their possible consenting and dissenting views on early SRM research. There is little to no identifiable commitment to establish substantive fora or events for Indigenous peoples to engage with others about whether such research should be conducted in the first place and, if so, what to research and how to conduct empirical inquiries. Policymakers, experts, and private citizens of the developed world[2] have a heavy moral burden to bear if they progress toward early SRM research without engaging in consent processes with Indigenous peoples.

I begin in the next section by claiming that the (arguably dominant) lesser of two evils argument for early SRM research can be construed as invalidating any potential dissenting views of Indigenous peoples. I deepen this claim by showing how this argument resembles an argument that has been used throughout history to silence Indigenous peoples from meaningful consent or dissent. I then move on in the following section to cover the scant literature that suggests possible consent processes for early SRM research. The common theme in this literature is that any fora or events for convening Indigenous peoples regarding SRM research should occur after research has been planned and even begun—thereby defeating the purpose of consent processes altogether. Consent or dissent after the fact is meaningless. In the concluding section, I argue that consent processes acceptable to Indigenous peoples must be based on partnership and include the following two requirements. First, Indigenous peoples should contribute actively to conversations about how to structure the consent processes in which they would participate. Second, in their interactions with Indigenous peoples, proponents of early SRM research are responsible for addressing them as sovereigns of their territories—despite the colonial conditions in many nations that frustrate Indigenous peoples' political independence.

THE LESSER OF TWO EVILS ARGUMENT

SRM involves techniques that decrease the absorption of solar energy by the earth and its atmosphere. Techniques include adding "light-scattering aerosols to the upper atmosphere" and "increasing the lifetime and reflectivity of low-altitude clouds" (Keith et al., 2010, p. 426). If implemented, the global effects of such SRM solutions will in some fashion impact everyone. Reports on geoengineering consistently argue that there is an urgent need for more research (Morgan & Ricke, 2010; U.S. Government Accountability Office, 2010). The question is why. Why should early research on SRM receive the support of the global public and funding from nations, corporations, Indigenous peoples, and organizations?

Stephen Gardiner summarizes well the essence of the commonplace argument for funding early SRM or other geoengineering research:

1. Reducing global emissions is by far the best way to address climate change.
2. In the last fifteen years or so, there has been little progress on reducing emissions.
3. There is little reason to think that this will change in the near future.
4. If very substantial progress on emissions reduction is not made soon, then at some point we may end up facing a choice between allowing catastrophic impacts to occur or engaging in geoengineering.
5. These are both bad options.

6. But geoengineering is less bad.
7. Therefore, if we are forced to choose, we should choose geoengineering.
8. But if we do not start to do serious scientific research on geoengineering options soon, then we will not be in a position to choose it should the above scenario arise.
9. Therefore, we need to start doing such research now. (Gardiner, 2010, p. 290)

This argument is a lesser of two evils argument as it hinges on the idea that doing research is desirable because it will allow human societies to avoid the evil of climate catastrophe at the expense of a less evil option: SRM and other geoengineering technologies. Even if the ultimate result of research does not achieve the most optimal conditions, they are still better than the conditions of a climate catastrophe. Though Gardiner suggests certain problems with this argument (including that it excuses moral corruption and depicts falsely the necessity of the two scenarios), these problems may not in the end overpower it (Preston, 2011). Instead of reviewing that particular discussion, I want to highlight one possible way the argument could be construed by its advocates in relation to Indigenous peoples.

Suppose that large numbers of Indigenous peoples reject SRM solutions because of their perceptions about negative impacts (including decreases in commitments to climate abatement) and their concern about the ideology of human control over the nonhuman world expressed by technologies like cloud whitening. Advocates of the lesser of two evils argument would acknowledge that these are reasonable responses to SRM. Advocates would also agree that SRM solutions are not ideal solutions. *But* given the likelihood of catastrophic climate changes, Indigenous peoples should shift their moral perspectives to what they would advocate for if the quality of *all* human lives were seriously in peril. Faced with such circumstances, everyone, Indigenous or otherwise, would agree that SRM is far from ideal, but necessary (Preston, 2011).

Advocates would likely claim that any dissent by Indigenous peoples to early research would be understandable, but not reasonable—and certainly dismissible. Advocates could ignore dissenters because the latter are not respecting the situation of humanity. Moreover, advocates could counter-argue that perhaps Indigenous peoples would be able to salvage more of their traditions thanks to the implementation of SRM than they would if a climate catastrophe were to run wild. Members of Indigenous peoples would be wrongheaded, thus worthy of being ignored, if they were unable to even see that SRM may be the best policy option for protecting their preferred lifeways—and hence their autonomy. Overall, then, possible dissenting views by Indigenous peoples would be counteracted based on the greater importance of potential human peril *as well as* the paternalistic idea that Indigenous peoples would be thwarting a viable option for protecting their preferred lifeways.

I have just sketched a construal of the lesser of two evils argument that can be seen as a mechanism for silencing Indigenous peoples and, for that matter, any other dissenters. Whether some or many Indigenous peoples do in fact dissent from early SRM research, and regardless of their reasons why, their dissent may be precluded in advance from mattering to the argument's proponents. In the next section, I will explore how this argument resembles arguments that have been used against Indigenous peoples many times in the past.

WE HAVE SEEN THIS KIND OF ARGUMENT BEFORE

Arguments of this kind are *not* new to Indigenous peoples, though they come in quite a few versions. In fact, they often mask a larger problematic. From many Indigenous perspectives, the "lesser of two evils" argument and the accompanying problematic look something like this. The problematic: non-Indigenous peoples cannot control their economies enough to stop a catastrophe or some other outcome perceived to be inevitable. The argument: Indigenous peoples should settle either for a meaningless role in determining the solutions for how to deal with such an impending outcome, or for being subject to feigned listening—followed by rejection of their views—and business-as-usual.

Consider a historical version of this argument that backed the building of twenty-nine major dams on the Columbia River during the twentieth century. The dams provide "flood control, navigation, and recreation benefits" (Federal Columbia River Power System, 2001, p. 5). In 1948, flood control in particular became a major priority after a twenty day long spring flood that took lives, destroyed one community, and wreaked substantial damage to property values (Northwest Power and Conservation Council, 2010). These rationales were argued for under the auspices that rapid U.S. economic growth required the dams (Barber, 2005; Harden, 1997; White, 1995; Ulrich, 2007). Moreover, Cold War anxieties demanded the United States do whatever it could to remain competitive and secure against the threat of global Communist political and economic hegemony. The dams promised power for the aluminum industry and hence the production of airplanes that would deliver atomic bombs. The plutonium would come from the nearby Hanford Engineering Works project (Reinhardt, 2011).

Dam proponents used the impending military threats and the inevitability of economic growth to silence dissenting views of tribal communities and small towns that were washed out by the reservoirs (Reinhardt, 2011; Barber, 2005). Tribal communities were concerned about the loss of their fisheries and land due to the reservoirs created by the dams. But their dissent was not judged to be meaningful. During public meetings about the construction of the Dalles Dam in the 1950s, for example, representatives of the Army Corps of Engineers did not offer any alternative solutions, did not consider any ideas that did not implicate the necessity of the dam, and failed to discuss the concerns or possible knowledge of the Indigenous community members. Barber recounts one tribal member's description of the meetings: "most of the talks made were by the agencies, which seemed to me to be like a group of friends, with a pat on the back—'You might say this,' and so on" (Barber, 2005, p. 69).

Dam propaganda dismissed the need for Indigenous peoples to continue their preferred lifeways. The Bonneville Power Administration film (1952) *Look to the River* claimed that these lifeways were "outdated and fruitless," yet dam development was "energetic and wise" (Cone, 2005). Moreover, on the day the Dalles Dam was completed in 1957, Senator Richard Neuberger reminded the crowd "our Indian friends deserve from us a profound and heartfelt salute of appreciation. . . . They contributed to its erection a great donation—surrender of the only way of life which some of them knew" (Barber 2005, p. 69).

The argument that justified the destruction of the Indian fisheries in the Pacific Northwest hinges on the idea that Indigenous dissent was not meaningful because of the urgency of the dams. This is a version of the lesser of two evils argument. Either the fishery gets saved and economic losses and flooding become rampant along with the corresponding threat to national security, or the fishery gets destroyed and economic growth ensues, flooding is reduced, and any serious threats are averted. The second option was judged to be the lesser of two evils. For dam proponents, the argument served to justify the silencing of dissenting views, which actually extended beyond Indigenous peoples to any others who depended on the fishery or whose communities would be submerged by the dams' reservoirs.

Both the arguments in support of dams and early SRM research derive their force from the anxieties over impending threats as judged by a certain set of experts, which in turn silences dissenting views. Both also presume there to be a single evil at stake, instead of many evils that will be suffered differently by diverse groups of people. The "evil" of fishery destruction actually refers to many possible evils that victims experienced in varying combinations and severities depending on whether they were white communities or tribal communities, for example. The same would be the case with any of the evils associated with SRM. And, finally, both arguments concern major technological endeavors. Many Indigenous peoples are accustomed to interpreting these kinds of arguments as cases of colonizers convincing themselves of the inevitability of some outcome as an excuse to dominate other people. Gardiner cautions us to "be wary of arguments from emergency; clearly, they are open to manipulation" (2010, p. 291). I would add to Gardiner's point that we should be even warier still of such arguments that fail to address the conditions perpetrating the emergency and that do not extend discussion to those with the most to lose.

It might be claimed (incorrectly) that the arguments in question are actually somewhat different because the Indigenous peoples were required to accept something that clearly would end their traditional lifeways, whereas that is certainly not the case made by proponents of early SRM research. In fact, some believe that the deployment of SRM could help some Indigenous peoples maintain their preferred lifeways in the event of a climate catastrophe. Hence, the two arguments are disanalogous when comparing the ultimate effects of the technological endeavors on Indigenous peoples' preferred lifestyles. This observation misses what creates the analogy between the dam arguments and the early research on SRM arguments. They are both fueled by an anxiety over wide-scale impacts that authorizes decision-making power over and against all dissenting voices. For example, the threat to national security partly motivated the building of dams, and atomic war was certainly understood as threatening the lives of anyone living within the United States, Indigenous peoples included. The protection of Indigenous peoples' lifestyles, in the sense of modernizing them, was implied in the Army Corps of Engineer's rhetoric of Cold War anxiety. This anxiety over wide-scale impacts is precisely the mechanism that authorizes decision-making power against presumably any would-be dissenters. This leads to a more serious similarity: in both arguments, decisions about whether to protect Indigenous lifeways and how they should be improved are placed solely in the hands of non-Indigenous peoples as a result of anxiety over wide-scale environmental impacts. That is, due to real or imagined anxiety, members of the dominant society determine the ultimate technological visions and priorities for the world. A lesser of two evils argument can use anxiety to sidestep morally necessary deliberative and consent processes with affected parties for visioning and setting priorities. Regarding this issue, there is little difference between early SRM research and the damming of the Columbia River.

The arguments are also similar to the degree that they exclude Indigenous peoples' visions of the future from consideration and secure non-Indigenous peoples' decision-making authority over traditional lifeways. They are also based on some conception of the unbridled growth of non-Indigenous economies. "Immediacy," "inevitability," "urgency," and "catastrophe" are code words for leaders and experts of dominant societies and corporations to decide *for* Indigenous peoples and others. Of course, advocates for early SRM research would deny that they are doing any of this and suggest that there are proposals out there for respecting Indigenous peoples' participation and engagement in early SRM research. I will turn to these in the next section.

A TANGLED WEB: APPROACHES TO CONSENT

Arguments from catastrophe and inevitability can serve to justify ignoring Indigenous peoples' visions and priorities for their own lives. Are there approaches to consent processes that could make such arguments less open to manipulation? I will outline some candidates and show how they feature certain problematic assumptions. That is, the approaches can effectively exclude Indigenous peoples altogether by suggesting they be consulted only after research plans have crystallized or by treating them as stakeholders but not as sovereigns of their territories. Indigenous peoples cannot accept these assumptions for *at least* this reason: early SRM research, because it represents the possibility of an *additional* intervention into the climate system, will be part of a process whereby members of dominant societies could increase their influence and control over Indigenous peoples' political and cultural systems for meeting basic needs and fulfilling preferred lifeways.

The first approach, which I call the *standards approach*, sees the concerns of affected parties as being met through setting appropriate limits and safeguards on research. The standards approach proposes that the possible nations and organizations likely to develop SRM research should develop formal standards and quality assurances that will be acceptable to stakeholders such as Indigenous peoples and the global public in general after research has already been planned and begun. A good example is Keith, Parson, and Morgan's "bottom-up" approach to international governance:

> A better approach would be to build international cooperation and norms from the bottom up, as knowledge and experience develop—as happened, for example, with the landmine treaty, which emerged from action by non-governmental organizations (NGOs). A first step might be a transparent, loosely coordinated international programme supporting research and risk assessments by multiple independent teams. Simultaneously, informal consultations on risk assessment, acceptability, regulation and governance could engage broad groups of experts and stakeholders such as former government officials and NGO leaders. Iterative links between emerging governance and ongoing scientific and technical research would be the core of this bottom up approach.[3]
> (Keith et al., 2010, p. 427)

Keith et al.'s governance model essentially begins with a core set of experts actually initiating research first, allowing then for other experts and stakeholders to provide wide advisory inputs on the standards that should be abided by. Standards of ethics, research integrity, and management of risks would be developed by the parties who would be constrained by them and who see value in having standards in the first place. The core idea is that *ethically acceptable standards and quality assurances for governing SRM R&D should be developed for or applied to loosely coordinated research.*

Indigenous peoples are not mentioned explicitly in this approach and could only be included as the stakeholders who would lodge comments and, possibly, complaints once research efforts were already underway. Indigenous peoples would not be considered as having inputs into the vision and priorities behind the research or even in any of the technical aspects of the research itself. But Keith et al. would likely claim that bottom up governance would at least assure Indigenous peoples that their safety concerns are being taken seriously. Parthasarathy et al. (2010) question whether Keith et al.'s bottom-up governance model really deals with the consent of affected parties in early research on SRM. They argue "that before the U.S. government funds further SRM research, or sanctions deployment of these technologies over the Arctic, it must sponsor deliberative democratic efforts that will engage the multiple populations across the globe who are likely to be affected" (Parthasarathy et al., 2010, p. 4).[4]

Parthasarathy et al. put forward an alternative strategy, which I call the *integrationist approach*. According to Parthasarathy et al., local communities and Indigenous peoples should "contribute their local knowledge in a deliberative manner at the grassroots level" (p. 7) to early SRM research. Indigenous peoples should be integrated in SRM research as a means of gaining informed consent. They claim that there are a variety of such approaches that have gained the informed consent of these stakeholders in contexts of other environmental issues. They claim these integration strategies influence "binding policy" (Parthasarathy et al., 2010, p. 4), resulting in a win/win situation: local communities like Indigenous peoples get to "participate in decision-making that has a direct impact on them, their homes, and their livelihoods . . ."; decision makers, then, get to "make direct use of indigenous knowledge, which provides one of the best measures of the changes taking place in the Arctic region" (Parthasarathy et al., 2010, p. 4).

Parthasarathy et al. give six examples of successful integration strategies which served as consent processes, grouped in pairs. The first two cases involve tribes who used the resources of non-governmental organizations (such as technical assistance) to strike negotiated agreements with Indonesian and Thai policymakers. The subsequent agreements better balanced each party's intentions for lands that were initially subject to intrusive national park and forest management systems. The second two cases involve policymakers who used methods that engaged stakeholders' concerns and knowledge regarding an Australian soil conservation program and the creation of an Austrian national park. The methods involved establishing both networks of community groups that shared technical information and planning committees that provided local stakeholders with opportunities for bringing their knowledge to bear on the decisions. The final two cases involve deliberative polling on funding infrastructure projects in China and participatory budgeting in Brazil. Based on these cases, the core idea is that *open, collaborative committees and other participatory mechanisms should be integrated with early SRM research as a morally acceptable way of engaging Indigenous peoples and other stakeholders.*

On this view, Indigenous peoples get to be integrated into the research after its basic trajectory has been determined, which yields benefits for all parties. Parthasarathy et al. are absolutely right to highlight the contributions Indigenous peoples can make to technical research and policy (e.g., traditional ecological knowledge). However, though Indigenous peoples would be included at the scale of local decisions about how to implement research, they would not participate at the scale of decisions about what vision and priorities are being attempted to be fulfilled by early SRM research. Parthasarathy et al. also suggest approaches that are silent on the sovereignty of Indigenous peoples over their territories. Rather, Indigenous participation is framed as entirely a matter of the provision of knowledge to others and the expression of their concerns as a best practice for facilitating policy decisions and government programs.

The disregard for sovereignty may be well-intentioned. Perhaps Parthasarathy et al. valorize the processes they describe because they are alternatives that may be used in colonial circumstances where respect for Indigenous peoples' sovereignty cannot be achieved in the short term. However, if this is true for the cases themselves, this cannot be an excuse for proponents of early SRM research. There is no reason why a consent process *fully inclusive* of leaders of Indigenous governments, participants in groups such as International Indigenous Peoples Forum on Climate Change and the Environmental Justice and Climate Change Initiative (among other projects/organizations like them), and Indigenous intergovernmental organizations such as the National Congress of American Indians in the United States cannot be part of visioning and priority setting for early SRM research. Moreover, the scientists who favor early SRM research cannot claim that such a process is impractical. This raises the question: impractical *for whom*? Non-Indigenous control over early SRM research as well as control over possible SRM deployment is obviously not very practical for Indigenous peoples because they would have to react to plans that they played little to no role in bringing about. Exclusion from decision making is especially impractical for leaders of Indigenous governments and organizations (including Indigenous intergovernmental organizations) because it makes it more difficult for them to carry out their responsibilities to the communities they represent. Nobody can appeal to practicality as a way of privileging non-Indigenous authority over Indigenous peoples without a clear argument for why non-Indigenous practicality matters more.

Finally, some of the cases mentioned by Parthasarathy et al. are referred to as co- or joint- management and have been around for some time in the literature on international development. There are certain known problems. It is not so fair to write, as Parthasarathy et al. do, that "[T]hese approaches provide evidence of successful negotiation between government authorities and local and indigenous communities over scientifically, economically, and socially complex issues" (p. 11). This conclusion does not reflect the complexity of those circumstances. For example, co-management regimes often proceed from and/or create bureaucracies that force Indigenous participants into discriminatory institutional patterns and procedures. They are also subject to forms of deliberative discrimination such as group polarization and hidden profiles that prevent Indigenous voices from playing a meaningful role, especially when it comes to Indigenous ideas about environmental guardianship and caretaking. Although Parthasarathy et al. are confident that co-management is a possible solution, a large literature exists that describes the daunting challenges (Nadasdy, 2005).

Whereas Parthasarathy et al. have clear advantages over Keith et al., they both suffer from similar drawbacks in relation to Indigenous peoples. Indigenous people are engaged and participate only *after* technical plans and visions for SRM research have crystallized. The sovereignty of Indigenous peoples is almost entirely ignored. The approaches assume that non-Indigenous peoples are in the position to dictate the visions and priorities from which the research plans proceed, and this presumption is often fueled by anxiety.

Some might argue that I am being too demanding here. It is inevitable that governments, corporations, and private citizens of nation-states will initiate projects without consulting Indigenous peoples. Keith et al. and Parthasarathy et al. provide reasonable practices for how to secure consent. But this criticism cannot make sense if we are talking about practices that will wrest away large amounts of control over Indigenous peoples' lands at the climate and landscape scales. Just as Keith et al. worry that rogue states will initiate their own SRM programs, so too are Indigenous peoples concerned that they will be controlled by whoever controls SRM. Indigenous peoples' consent cannot be cheapened through appeals to impracticality and standard historical practice when we are talking about technologies that could possibly further threaten Indigenous political and cultural systems for meeting their basic needs and fulfilling their preferred lifeways.[5]

CONCLUSION: A BETTER APPROACH

A better approach to consent is not found in the literature on SRM but elsewhere in literature that covers free, prior, and informed consent (FPIC). Early research that risks placing large degrees of control over Indigenous peoples' basic needs and preferred lifeways must be bound up with an earlier consent process that accords with Indigenous peoples' customary laws and decision-making processes, for example, the United Nations Declaration of the Rights of Indigenous Peoples (UNDRIP) and other international human rights treaty bodies (Colchester & MacKay, 2004; MacKay, 2004). To argue *against* this would be to argue that non-Indigenous members of developed and developing nations have the privileged role in determining the vision and priorities *for* Indigenous peoples. This position cannot be sustained without the permission of specific Indigenous peoples or a diverse group of representatives of Indigenous governments, other community leaders, and Indigenous intergovernmental organizations. Any process that eventually secures Indigenous peoples' consent must have been structured collaboratively (i.e., in the spirit of partnership) so as to genuinely include Indigenous peoples (i.e., FPIC) customary laws and decision-making processes. As distinguished from the standards and integrationist approaches, I will refer to this approach as the *partnership approach*.

For many Indigenous peoples, FPIC implies respect for their sovereignty over the territories that they depend on for their basic needs and for fulfilling their preferred lifeways.[6] As customary laws and decision-making processes flow from the political and cultural systems that issue and legitimate them, another requirement for consent, then, is that Indigenous peoples must be acknowledged as sovereigns of their territories. Acknowledgment is required despite the colonial conditions in many nations that frustrate Indigenous peoples' political independence or have weakened their governing capacities.

In the case of early SRM research, nothing less than a consent process that meets the requirements of (1) FPIC and (2) acknowledgment of Indigenous peoples' sovereignty can provide ethical consent. A lesser of two evils argument for early research on SRM does not meet these requirements because it cannot justify why, except through an appeal to anxiety, even in times of emergency, non-Indigenous peoples get to set the vision and priorities for Indigenous peoples. Additionally, it is only after (1) and (2) have been met that the important features of the integrationist approach (e.g., traditional ecological knowledge) suggested by Parthasarathy et al. may be explored respectfully to Indigenous peoples in relation to SRM.

With the momentum toward SRM research already established, it is at best unclear whether things will go any differently for Indigenous peoples than in previous technological endeavors. Perhaps all I have done here is to highlight some of the reasons why proponents of early SRM research have not done enough to respect members and leaders of Indigenous peoples. Perhaps, as has been the case many times before, protest will ensue, and the ingenious political maneuvering of Indigenous leaders will help to counteract the inevitable dominance of others over their communities' affairs and future possibilities.[7]

NOTES

1. Indigenous peoples are human groups the members of which descend directly from pre-invasion inhabitants of lands currently dominated by others, such as Tribes and First Nations within the United States and Canada and Aboriginal and Torres Island communities within Australia. My definition Indigenous peoples is informed by other current definitions (See Anaya, 2004, 3; UNESC, 1994, Article 7; Metcalf, 2004; ILO, 1989).

2. I might add the developing world too insofar as developing countries also dominate Indigenous peoples.

3. The claims just stated resonate with those of many others who are concerned with the governance of early SRM and other geoengineering research in terms of staving off public resistance (House of Commons, 2010; Giles, 2010).

4. This working paper published online has no pagination. I followed the pagination of the PDF.

5. The 2011 Solar Radiation Management and Governance Initiative (SRMGI) report, convened by the Environmental Defense Fund, the Royal Society, and the Academy of Sciences for the Developing World (TWAS, based in Trieste, Italy), has some of the same problems featured above. Indigenous peoples are simply not mentioned at all. In fact, in this report, SRM research types are put into categories and the report determines that only some of the categories require stakeholder and public engagement. I can only hope that there will be an opportunity for this report to be subject to transformation by Indigenous peoples, as the report gives no evidence that there was adequate participation of Indigenous peoples in its generation.

6. There is some worry that international law, including the UN Declaration on the Rights of Indigenous Peoples (UNDRIP), does not strongly represent Indigenous peoples' sovereignty. My claim here is that FPIC, however, implies that sovereignty is respected. A partnership approach would include all of the attractive features of international law concerning Indigenous peoples *and* also promote sovereignty owing to the requirements of FPIC.

7. Special thanks are due to Larry Merculieff, Kristie Dotson, Robert Figueroa, and Christopher Preston for their insightful comments on earlier drafts.

Chapter Five

Solar Radiation Management and Vulnerable Populations

The Moral Deficit and its Prospects

Christopher J. Preston

> SRM research could constitute a cheap fix to a problem created by developed countries, while further transferring environmental risk to the poorest countries and the most vulnerable people. —SRMGI, 2011, p. 21

> Populations living at the edge of subsistence—those with the least capacity to adapt to the impacts of climate change and almost no voice in international deliberations—are precisely the populations that will be most vulnerable to any negative side effects that geoengineering experiments may have. . . . Do the populations most vulnerable to harm by geoengineering experiments deserve some opportunity for informed consent? And if so, how can this be done? —Olson, 2011, p. 39

From its inception in 1992, the United Nations Framework Convention on Climate Change (UNFCCC) has recognized that countries differ with respect to their vulnerability to the effects of climate change, their historic responsibility for causing it, and their economic capacity to adapt to its harms. As a result of these differences, parties to the original convention agreed to safeguard the climate system "on the basis of equity and in accordance with . . . common *but differentiated* [emphasis added] responsibilities and respective capabilities" (UNFCCC, 1992, p. 4). Commentators from Bill McKibben to Al Gore to the Royal Society have all acknowledged that anthropogenic climate change creates a moral burden. The fact that the UN in 1992 declared the burden "*differentiated*" was acknowledgment that responsibilities fell differently on different parties. The framework convention's 1992 distinction between "Annex I," "Annex II," and "non-Annex" countries was a crude attempt to divide these responsibilities among signatories.

Although the idea of differential moral burdens is integral to the framework, the history of the Conference of the Parties (COP) meetings since 1992 suggests that signatories have never managed to fully agree on how these responsibilities would be divided up in practice.[1] If and when climate engineering is added to the package of responses to global warming, these distributional differences are likely to become even more complicated. It is becoming increasingly clear that differential impacts and responsibilities of climate change have the potential to be exacerbated by differential impacts and responsibilities in the light of future climate engineering. It is important, then, at the start of the era of serious research into climate engineering, that we consider how these differential moral burdens might play out on the path ahead.

When the UN convention was adopted in 1992, climate engineering was not yet viewed as a serious option for mitigating the effects of global warming. This changed in 2006 when Nobel Laureate Paul Crutzen wrote an influential paper suggesting that, while emissions reductions were by far the best response to anthropogenic warming, the lack of action on emissions meant that intentional climate modification with stratospheric aerosols should be seriously researched (Crutzen, 2006). Crutzen did not advocate stratospheric aerosols with any enthusiasm. But after reflecting on the world's track record in the fourteen years since the UNFCCC had been adopted, he felt little hope that the global community would succeed with any of the better options.

The belief that responding to the climate problem would involve technological innovation has always been a key component of the UN approach. Without talking about specific technologies, the original convention encouraged technological transfer from developed to developing countries, calling on signatories to ". . . promote and cooperate in the development, application and diffusion, including transfer, of technologies, practices and processes that control, reduce or prevent anthropogenic emissions of greenhouse gases" (UNFCCC, 1992, p. 5). Though it is unlikely that SRM is the type of technology the authors of the 1992 document had in mind for addressing the problem of excessive emissions, the technology is rapidly gaining currency as a possible response to climate change. The recent inclusion of two days of geoengineering discussion at the June 2011 Intergovernmental Panel on Climate Change (IPCC) meeting in Lima, Peru, and the inclusion of evaluations of geoengineering in the IPCC's 5th assessment report due out in 2013–2014, leave little doubt that SRM is now being considered as a possible response to warming temperatures.

As most of those engaged in this discussion have conceded (Crutzen, 2006; Robock, 2008, Corner & Pidgeon, 2010), the possibility of SRM raises numerous difficult moral concerns. For example, in its atmospheric forms, SRM would operate on what is normally considered a global commons. Its benefits and burdens are likely to be distributed unevenly and uncertainly. In the course of creating cooling, some forms of climate engineering could lead to—or ensure the continuation of—problems such as ozone depletion, changes in precipitation patterns, ocean acidification, and the disruption of terrestrial ecosystems. Implementation could be multinational or unilateral, well-intentioned or malicious. Even in the rosiest of scenarios, successful implementation of SRM would demand a hard-to-achieve international commitment over an unspecified period into the future. Even embarking on this journey could influence the motivation of nations to pursue still needed emissions reductions. SRM clearly adds a number of new and difficult moral challenges to climate change.

Among all of these challenges, concerns about global justice loom particularly large. As well as potential benefits, SRM promises substantial new threats to global equity. When the threats engendered by SRM are layered on top of the differential responsibilities engendered by unintentional warming, a glaring "moral deficit" between rich and poor countries has the potential to be worsened.[2] Taking this moral deficit seriously seems to be a prerequisite for a globally just and fair approach to geoengineering. My purpose in what follows is to briefly articulate how the imbalance already existing in anthropogenic warming is potentially aggravated by SRM. Then, I will make some tentative suggestions about a helpful moral lens that can be used to view this deficit. Finally, I will discuss some initial measures to take in the face of this deficit as research and discussion of potential deployment of SRM gathers pace.

SKEWED VULNERABILITIES AND COMPOUND INJUSTICE

In 1992, before the discussion of geoengineering had got off the ground, Henry Shue warned about the possibility of "compound injustice" in international negotiations over climate change (Shue, 1992). What Shue meant by compound injustice was that past inequalities in international relations had weakened the ability of vulnerable nations to achieve fair treatment in climate negotiations. Less developed nations had already lost the economic lottery through colonialism and military and economic imperialism. Now they stood to lose again due to their political weakness in discussions of how to proceed on global warming. Already present in anthropogenic warming, Shue's worry about compound injustice rears its ugly head again in the debate over geoengineering. In fact, the compounding of injustice has the potential to get worse. Seeing how this injustice might accumulate is the first step in determining what should be done about it.

a) *Geographically increased vulnerability to climate change.* It is well known by now that climate change promises to wreak the greatest havoc and destruction on the lives of the global poor partly as a result of nothing more than geographical bad luck. Many of the global poor happen to live in locales that will be most susceptible to the consequences of increasing global temperatures. In Bangladesh, for example, with an average per capita income of $1700 p.a. (adjusted dollars for purchasing power parity), a one meter rise in sea level would inundate 18% of the land area, home to 11% of the population (over 16 million people) (Argrawala et al., 2003). At the same time, a decrease in the runoff from shrinking glaciers is projected to lead to up to one billion people facing water shortage and subsequent land degradation throughout Asia (IPCC, 2007b). In a similar cruel piece of geographical misfortune, one third of African people live in drought prone areas. The 2007 IPCC report predicts that climate change will result in the loss of significant agricultural lands in Africa due to more extreme weather, causing shorter growing seasons and lower yields. With temperatures in Africa predicted to rise up to 4 degrees Celsuis over the next century, the geographical range of diseases such as malaria, tuberculosis, and cholera may also increase.

b) *Economically increased vulnerability to climate change.* Situated squarely on top of the geographical bad luck is the fact that persons who lack resources and economic mobility are less capable of extricating themselves from life-threatening situations. As Hurricane Katrina amply demonstrated, without adequate individual wealth to insulate oneself, the seriousness of natural calamities quickly escalates even in one of the richest countries on earth. Throughout much of Africa, according to the United Nations, a number of factors, including poverty, illiteracy and lack of skills, weak institutions, limited infrastructure, lack of technology and information, low levels of primary education and healthcare, poor access to resources, low management capabilities, and the prevalence of armed conflict all have "negative effects on the continent's ability to cope with climate change" (UNFCCC, 2007, p. 18). Cash-strapped and debt-burdened governments are less able to provide the necessary services to alleviate the human suffering that drought, flooding, and refugee crises will cause (IPCC, 2007c).

c) *Historical responsibility.* The geographical and economic vulnerability of the poor nations to rising temperatures is particularly unfortunate given their lack of historical responsibility for creating the problem in the first place. According to World Resources Institute figures, from 1850 to 2000, just 25 countries accounted for 90% of all carbon dioxide emissions (Baumert et al., 2005). Due to the different periods in which particular countries industrialized their economies, a country's historic share of global emissions might exceed their current share. The United Kingdom, for example, whose emissions in 2008 amounted to just 1.73% of the global total, is responsible for 6.3% of global historical emissions due to early fossil fuel use during its industrial heyday.

The three factors described illustrate how, through a combination of skewed vulnerabilities and skewed responsibilities, climate change appears to be particularly unfair to the poorer nations. As Chris Cuomo has put it, "[c]limate change was manufactured in a crucible of inequality" (Cuomo, 2011, p. 693). Since the poorer countries appear to be geographically and economically most in need of relief from the effects of climate change, one might expect them to have the most invested in the cooling that SRM aims to achieve.[3] The skewed vulnerabilities to climate change suggest skewed interests in avoiding it. While SRM-induced cooling certainly has the potential to help the most vulnerable, several considerations make the technology considerably less appealing to the vulnerable nations than it might at first appear.

d) *Geographically increased vulnerability to the uncertainty in SRM.* Because SRM will reduce the amount of solar energy entering the climate system, it will also alter precipitation patterns in ways that are not entirely predictable. Robock et al. (2008) have argued that both tropical and Arctic injections of stratospheric aerosols could disrupt African and Asian monsoons. Trenberth and Dai (2007) found a similar likely decrease in precipitation patterns modeled on the effects of past volcanic eruptions. Bala et al. (2010) concluded, in contrast, that marine cloud brightening could lead to wetter continents. Matthews and Caldeira (2007) projected a variety of impacts on precipitation in Africa, Asia, and South America due to changes in rates of evapotranspiration in vegetated regions. The uncertainty inherent in predicting precipitation (due in part to complicated water vapor feedback loops) means that the precise consequences of SRM on the hydrological cycle are unclear. While SRM may allow crop productivity to be maintained (or even enhanced) overall in areas experiencing stress from climate related warming (Pongratz et al., 2012), significant regional variability means that SRM still may "pose a risk to local food security if subsistence farming prevails and adaptation is not possible" (Pongratz et al., 2012, p. 3). So, paradoxically, some of the same populations that are the most geographically vulnerable to the effects of climate change also appear to be geographically vulnerable to some of the uncertainties inherent in the geoengineering that might save them.[4]

e) *Economic vulnerability to SRM.* A parallel argument about economical vulnerability can also be constructed for the uncertain effects of SRM. Just as impoverished nations are ill-equipped to deal with the hardships of climate change to which they are disproportionately exposed, they are similarly ill-equipped to deal with the uncertainty inherent in the effects of SRM. As Keith, Parson, and Morgan (2010) have stressed in a paper in *Nature*, "it is vital to remember that a world cooled by managing sunlight will not be the same as one cooled by lowering emissions" (p. 426). If monsoon patterns change and rainfall either decreases or increases dramatically, the disruption will present far greater challenges to those countries that lack adequate economic resources than to those countries that are more economically prosperous. Effects on precipitation, especially increases in extreme weather events, will impact developing world farmers and those who live subsistence lifestyles the most. Other challenges also await the poor. A planet successfully cooled with SRM but with high CO_2 levels in the atmosphere is unlikely to provide the marine resources upon which the populations of some less developed nations depend for subsistence, due to greatly increased acidity levels in the oceans.

f) *Lack of influence in development of SRM technologies.* It comes as no surprise that research into SRM is taking place predominantly in the developed countries. Only those countries with a certain type of research infrastructure are likely to have the resources to study, simulate, and field test SRM. The United States, the UK, and Germany are among the nations with the largest number of scientists currently researching SRM. China and India are rare exceptions among non-annex countries for engaging in some geoengineering research. This predominance of scientists from developed nations means that the most vulnerable people are unlikely to have much input into the shape of SRM research. This introduces the possibility that the interests of the developing countries will be less well served by prospective geoengineering than they might be if the science of SRM were controlled more equitably. One might reasonably assume that scientists in the richer countries will have the interests of the most vulnerable in mind as they develop and test particular SRM strategies. Despite good intentions, this may not be how matters ultimately play out. There is a preponderance of examples from the history of science of allegedly neutral research being dramatically (and often unwittingly) distorted by self-interest (Gould, 1981; Bleir, 1984).

g) *Political vulnerability.* When these observations about the location of SRM research and development are coupled with glaring imbalances of power in international decision-making structures (as detailed by Shue and others), it begins to look highly likely that the poorer nations will not be treated as equal participants in decision making over the implementation of SRM. The COP meetings of the UNFCCC have clearly demonstrated over the years how the interests of vulnerable nations are repeatedly held hostage to the convenience of the more powerful voting blocs. In today's climate discussions, even a single high-emitting country can stymie the carbon reduction goals of a large collection of lower-emitting nations. Ricke et al. (2010) have shown that the interests of different regions in the amount of SRM desirable will increasingly diverge over time. In future SRM discussions, one wealthy developed country with advanced technological infrastructure could potentially dominate the discussion either through the threat of unilateral action or through the threat of veto power. The compounding of injustice that Shue worried about with climate change seems highly likely to reappear in future discussions of SRM research and deployment.

It should be clear from the above that the most vulnerable nations are not only suffering the greatest climate injustice today, in some cases they risk having those injustices compounded tomorrow as the development of geoengineering technologies and discussions about governance unfold.[5] Poor nations find themselves on the wrong end of a moral deficit of startling proportions, a deficit accumulating through the following:

a. Geographical vulnerability to climate change
b. Economic vulnerability to climate change
c. Lack of historical responsibility for climate change
d. Geographical vulnerability to the uncertainty in SRM
e. Economic vulnerability to the uncertainty in SRM
f. Lack of influence in the development of SRM technologies
g. Lack of influence in the future governance of SRM

The next section of the paper proposes one potential, if intentionally minimal, moral lens through which to view this deficit.

DESERT AND THE MORAL DEFICIT

Given the inequities accumulating, one might ask about the specific moral burden these imbalances impose on the rich countries. While it is beyond all doubt that the rich countries owe the poor countries a considerable amount for the injustice accrued (Shue, 2010; Singer, 2002; Gardiner, 2004), there are factors at play that make the apportioning of specific obligations both complicated and contentious. Rich countries do not possess their wealth for malicious reasons (even if poor countries have been—and continue to be—exploited during the accumulation of that wealth). Nor has the historically disproportionate contribution of the rich countries to climate change been ill-intended. Rather it originally served merely to help lift their own populations out of poverty. The geographical bad luck that exacerbates climate injustice for the poor nations is just that—bad luck. It may be difficult to identify which rich countries have concrete obligations to which poor countries as a result of mere luck. As climate negotiations have demonstrated, the whole idea of differentiated responsibilities has been controversial. When considering the additional imbalances that may accrue through SRM, it certainly seems unfair to accuse the well-meaning scientists who are investigating geoengineering in the developed nations of intentionally compounding injustice rather than trying to do something to help. More complexity is found in the fact that the richer countries are not economically homogeneous within their borders. The country with the greatest per capita emissions, the United States, also has substantial income disparities between rich and poor, making it unlikely that the moral burden falls equally on all citizens within that country.

While the point here is not to deny that specific obligations do fall on the rich countries, the complications suggest there may be reason to look for ways to supplement claims of direct obligation and responsibility for remedying the moral deficit with some different analysis. As a practical matter, existing talk about the moral obligations of the rich countries to the poor over climate injustice has yielded decidedly mixed—and certainly inadequate—results. The concepts of obligation and responsibility have not proven, on their own, to be capable of motivating the required action. Though critics would suggest that much of this inaction is the result of mere laziness and wanton immorality, some of it might also be for partially legitimate, epistemic reasons. There may be goods deserved by the poor countries for which responsibility is simply too difficult to assign. Questions about migration and relocation policies, the extent of technology transfer and compensation required, and how to treat economies with rapidly increasing carbon intensity are genuinely difficult to resolve. In addition to the epistemic challenges of correct apportionment, there may be further goods that rich countries have no specific obligation to provide even though the poor countries manifestly deserve them in the face of climate change (e.g., freedom from dictatorial rulers or a greater abundance of key natural resources). Given the uncertain future ahead in a warming world, the concepts of obligation and responsibility may be neither broad enough nor flexible enough for the demands that climate change and SRM will make. For these and other reasons, when investigating the potentially compounding injustices of future SRM, there may be a need for additional moral analysis. That analysis may reside partly in the idea of desert.

George Sher has claimed "desert is central to our pre-reflective thought" (Sher, 1987, p. ix). At first face, it appears that in many different areas of concern certain persons or groups *deserve* something and others do not. The careful pedestrian does not *deserve* to get hit by the drunk driver. The hardworking student *deserves* the good grades. The persistent job-seeker *deserves* a lucky break. The idea of desert comes into play in numerous situations, containing varying degrees of moral significance. At this pre-reflective level, it seems likely that the developing nations do not *deserve* the hand that climate change and the prospect of SRM deals them.

The concept of desert differs from the idea of obligations and rights. It is possible for someone (or for some group) to deserve something to which they have no rights. It is also possible for one group to deserve something when no other group has any particular obligation to provide it. Sher uses the example of a vicious person who inherited a fortune while those around her lack basic necessities. While the inheritor may be less deserving than the poor, those in need do not have a right to her fortune. Similarly, the lucky winner of the marathon whose opponent tripped and fell to the ground 10 meters from the line is under no obligation to hand over the medal even though the other runner might have deserved to win. In some respects, then, the normative force of desert appears to be weaker than the force of rights or obligations. Even when people are clearly deserving of something, there may be no specific individual with the obligation to provide it and no moral guarantee that they will get it.

This somewhat subdued moral power of desert relative to rights and obligations in no way means that desert lacks all moral force. According to Sher, the moral force of desert unpacks in different ways depending on the type of desert in question (compensation, punishment, reward, etc.). In the case of climate change and SRM, the most suitable unpacking in Sher's analysis is his suggestion that desert is sometimes an expression of what is required to restore diachronic fairness.[6] Sher states there are "a whole range of desert-claims that appear to require some trans-temporal balancing of benefits and burdens" (Sher, 1987, p. 91). Some of these burdens might be wrongfully inflicted (e.g., through economic imperialism) and some might be inadvertently inflicted (e.g., through geographical bad luck). Regardless of how they have come about, in the interests of diachronic fairness the bearer of the burden is deserving of some redress.

While the tie to diachronic fairness may initially suggest that the normative force of desert is based ultimately in a familiar principle of fairness and a concomitant obligation of some individual, country, or supranational body, we have already seen that it is possible for desert to exist without any obligation at all. In the absence of direct obligation, Sher points out that an alternative interpretation of diachronic fairness places the normative force in the desirability of restoring situations that contain more value than the present one. In the current case, this suggests a commitment to the belief that a world in which no countries are disproportionately burdened by climate impacts is a more valuable world than the current one. Various philosophical frameworks could be used to justify this claim. A broadly utilitarian framework would appeal to the total happiness at stake, a Rawlsian framework to matters of distributive justice. Whatever it is, this framework supplies an "independent standard" whose violation is the source of the burden that now needs balancing (p. 94).

Since their normative force rests only in a claim about some situation being more valuable than some other, it is striking that the demands of diachronic fairness do not necessarily pose an immediate obligation on anyone. At first, this might seem paralyzing and unhelpful. Indeed, Sher acknowledges that ". . . if desert-claims tell us only what has value, then desert may seem to have no bearing on how we should act" (Sher, 1987, p. 202). But he quickly counters this worry by suggesting that while desert might not impose any *direct* obligations, it may yet generate a wide range of *indirect* obligations that certain individuals or countries possess to make the desirable situation come about. "Even if desert-claims do not themselves dictate actions," Sher claims, "the values from which they draw their force may surely have an important influence on the obligations that do" (p. 202). If a different distribution of climate burdens is more valuable than the current one, then certain parties are going to find themselves with indirect obligations to make this more valuable situation come about.

Even though these obligations associated with redress are indirect, Sher leaves no doubt about the direction in which they point. "[A]ny obligations generated are apt to apply primarily to those who were implicated in the original violations [of the independent standards]" (Sher, 1987, p. 204). In this case, the original violation was perpetrated by those countries that exceeded their equitable share of the atmosphere's capacity to absorb greenhouse gases (*viz.* the rich countries). The exact nature of these indirect obligations will be highly contingent, hinging, Sher claims, on a number of factors. These factors include the urgency of the situation (e.g., How much harm is accruing? Is it imminent?), the comparative ease (or difficulty) with which the valuable situation could be promoted (e.g., How expensive is it or how much change in lifestyle is required?), prior commitments to promote these goods (e.g., What sort of international commitments to global justice exist in the world?), and the presence of other conflicting obligations (e.g., the obligation of a country to its own citizens) (p. 203). While Sher's discussion of desert and diachronic fairness originally had no connection to the climate issue, these turn out to be exactly the sorts of contingencies nations wrestle with in climate negotiations. The indirect—and somewhat less determinate—imperatives generated by desert seem to speak more faithfully to the situation that the rich countries find themselves in with regards to climate injustice. Not only does the analysis seem to ask many of the right questions, it also allows greater flexibility for future redress. The suggestion here is certainly not that desert can *replace* the concept of obligation in discussions of climate injustice. It may, however, provide a *valuable supplement*.

STARTING TO REPAY THE MORAL DEFICIT

Viewing climate injustice through the lens of desert opens the door to a broad range of actions toward vulnerable nations in the face of the potentially compounding injustices of SRM. This latitude is valuable given that future decisions about climate policy can draw on no good analogue from the past. The technologies, the politics, and the science of climate change are all dynamic enough that the moral deficit accrued might justify a range of different demands as technologies and events unfold. Some actions (e.g., monetary compensation, logistical assistance with adaptation) might indeed end up being direct obligations for the rich countries. Others (e.g., certain types of technology transfer, more permissive immigration policies toward ecological refugees, a greater role in decision-making) might be better explained as matters of desert. Because SRM policy is currently in its infancy, with no actual deployments imminent and most countries still remaining officially agnostic on the technology's desirability, there are only a few questions in relation to SRM that are actionable today. Of those that are, the question of how to proceed ethically with research and preliminary planning for governance of SRM is one of the most pressing.

Numerous recent reports, including the Royal Society (2009), Asilomar Scientific Organizing Committee (2010), Olson (2011), the Solar Radiation Management Governance Initiative (SRMGI) (2011), and the Bipartisan Policy Center (2011), have all emphasized the importance of public engagement in the early stages of geoengineering research. A proper type of public engagement provides more than just an opportunity for a research community to check a box and say that they have consulted with some citizens about what they are planning to do. True engagement views citizens as active participants in both technological design and policymaking (Wynne, 2006). The SRMGI report characterizes this type of public engagement as being about ". . . moving away from models of prediction and control toward a richer public discussion about the visions, ends and purposes of science and technology" (SRMGI, p. 40). The belief behind public engagement is that the ideas citizens contribute in discursive forums such as focus groups, town hall meetings, consensus conferences, and deliberative polls can lead not only to better technology policy but sometimes also to better technology. This type of engagement is particularly important for those technologies being designed to serve a large public policy goal with numerous social impacts. SRM is without doubt one of these.

The SRMGI report viewed public participation broadly, encouraging engagement with "the breadth of emerging stakeholders in the dialogue" (SRMGI, 2011, p. 10). The report saw the need for discussions to be "progressively broadened to include representatives of more countries and more sectors of society" (p. 10), adding "[s]ince climate change is a global problem research into possible SRM methods . . . should be open to global scrutiny" (p. 24). Unfortunately, while admirable in many respects for its inclusive tone, this call for an increasing breadth of engagement made no mention of the particular importance of engaging with vulnerable peoples, noting obliquely only the "considerable challenges" (p. 41) involved in engaging with publics in those countries in which little or no geoengineering research or discussion is currently taking place. Yet of all the populations that should participate in the early days of SRM research, it would appear that the most vulnerable populations in the poorest countries are uniquely deserving.

The argument for the special importance of the participation of vulnerable populations in early discussions of SRM research and policy can be made in a number of ways. Here I employ just two. There are both "normative" and "substantive" justifications (Stirling, 2005) for particular attention to the voices of the vulnerable in SRM research and development.[7] The normative argument for public engagement with the vulnerable is that it is simply the right thing to do. Stirling claims that such a view about engagement in general is often defended by appeals to justice or democracy through authorities such as Rawls or Habermas. The claims made in the previous section strongly suggest that, in this case, engaging with vulnerable populations in the early stage of SRM research is simply a matter of desert. A population (1) seriously threatened with climate dangers resulting from the actions of others, and (2) potentially further threatened by the actions of those attempting to correct the problem clearly deserves to contribute meaningfully to the development path of any SRM technology. While failing to mention vulnerable populations specifically, SRMGI appears to endorse this claim, arguing that if SRM really is intended to be a *public* good then "those responsible for overseeing the research need to make every effort to ensure that the public understands and agrees that it wants to pursue this option, and is consulted as inclusively as possible in decision-making processes throughout any research programme" (SRMGI, 2011, p. 22). The word "throughout" is significant. Waiting until field testing before inviting the participation of vulnerable populations will be too far downstream in the technology's development. Momentum will already be in place. The types of SRM being pursued, the methods, and their particular goals will likely be locked-in by a cadre of invested scientists and institutions. This would not be what the poor nations deserve.

Substantive justifications are more controversial, but reflect the idea that public engagement can actually enhance the quality of research thereby leading to improved technical and policy solutions. Stirling points out that this kind of engagement is particularly important when there exist "intractable scientific and technological uncertainties" (Stirling, 2005, p. 223). Under such conditions of uncertainty, engaging with stakeholders is important ". . . as a means to consider broader issues, questions, conditions causes or possibilities" (p. 223). Untested technologies with broad social impacts and inherent uncertainties demand the widest possible range of questions to be raised. Though he did not have SRM or any other geoengineering technology in mind, Stirling could have been referring directly to the climate situation when he pointed out that the advantages of "opening up" discussions about research through public engagement include the following:

> [C]onsidering more indirect (and wider ranges) of effects, extending timescales and geographical scope, examining different sensitivities and scenarios, scrutinizing benefits and justifications (alongside risks), comparing alternative choices and including strategic issues such as reversibility, flexibility, and diversity. (2005, p. 223)

Points made by Stirling about engagement in general are strikingly similar to more targeted points made about engagement with vulnerable peoples on the climate issue by Intemann (2011). Intemann draws attention to work in contemporary science studies that emphasizes methods for countering the presence of potentially harmful non-epistemic (i.e., social) values within research. The broad idea, she suggests, is "to structure scientific communities and practices in ways that encourage the identification and critical evaluation of background assumptions, theories, and models" (Intemann, 2011, p. 495). This can be achieved by methods such as including researchers with diverse experiences, values, and social positions, maximizing avenues for scrutiny of assumptions, and, most notably, investigating scientific phenomena from the perspective of those stakeholders that are often marginalized. These methodological prescriptions have clear applications for SRM research.

The SRMGI report breaks SRM research and development into five categories, moving progressively from work that takes place entirely in the laboratory on computers to work that involves large scale deployment. The report claims that as you move up the list of five categories toward deployment "there are progressively stronger grounds for informing and consulting the public" (SRMGI, 2011, p. 41). While few would doubt the need for participation of the vulnerable public at the level of decisions about large scale field tests and deployment, Intemann's position supports the idea that such participation is also beneficial at the very earliest stages of technological development. This is not simply because of the possibility of technological lock-in. Participation of the vulnerable has what Stirling would call substantive benefits, actually improving the science from the very beginning. According to Intemann, this is the case even when the research takes place at the level of computer modeling. While this is not the venue to make a full argument for the benefit of wide participation at the computer modeling stage of climate research, there is an important strategic reason to start the discussion here. If the argument for wide participation can be made with any plausibility at all at the level of climate modeling on computers, then the case for participation of vulnerable persons at all the later stages of research and development becomes increasingly compelling.[8]

Intemann adopts a position recently supported by Winsberg (2010) and suggests that even computer-based climate modeling in the lab can inadvertently introduce non-epistemic values. Given the complexity, non-linearity, and uncertainty in climate science, climate models invariably make assumptions about such things as what parameters to keep constant, what variables to measure and how to prioritize them against each other (e.g., the emphasis on global mean surface temperature rather than pressure or precipitation), what to use as starting conditions, what impact various feedback mechanisms will have, and how far in the future to run the model. If the model is predicting impacts, further assumptions about what to count as costs and what to count as benefits also come into play. The more social the impacts become, the more room there is for disagreement, both about what to count and about how to compare them. Some costs, such as GDP impacted and property damaged, lend themselves to (relatively) easy quantification and are more likely to be reported in climate impact models. Others, such as increased dependence on foreign aid, biodiversity loss, and loss of cultural traditions, are not so easy to include. Many of these costs are incommensurable, making it difficult to meaningfully compare the results of different scenarios.[9] These decisions about how to set up models, run them, and evaluate the results may be shaped by the predispositions of the researcher or the research community, together with the prevailing standards currently existing within the discipline. As Intemann puts it, "value judgments implicitly built into climate models may be unsupported or may, inadvertently, reflect only the interests of modelers" (Intemann, 2011, p. 498).

The recommended way of countering these non-epistemic values is to structure investigative communities in ways that offer a more critical evaluation of background assumptions, theories, and models. The 2007 IPCC assessment report emphasized the "crucial" role of stakeholder involvement in impact assessments (2007c, section 2.3). Intemann's recommendation about the value of investigating phenomena from the perspective of the marginalized has particular salience for the form of stakeholder involvement in climate issues. Studying "from the margins-out" has what she calls ethical and epistemic motivations (matching the normative and substantive justifications for public engagement detailed by Stirling). In the case of SRM research, the ethical motivation is that the interests of those most vulnerable to climate change and with no role in causing it simply deserve to be heard. The epistemic motivation is that the perspective of those who have so far been only on the receiving end is the perspective least likely to overlap with those who make up the majority of the scientific community doing the research. From this vulnerable perspective, problematic assumptions may be more apparent. This is particularly important when researching a technology designed to serve a policy goal as broad as cooling the whole planet. As Intemann puts it,

> [I]t will be important to involve researchers from different geographical areas (e.g., from developing countries in the Southern Hemisphere as well as from developed countries) who will be in a better position to evaluate whether climate models or policies address the needs and interests of populations in those areas. Researchers from developing counties, for example, would be better situated to realize when certain impacts that are salient to those in the global South are being neglected.... When the interests and needs of marginalized populations are not taken into account from the beginning, even well-intentioned science may lead to policies and interventions that reinforce rather than address existing social inequalities. (Intemann, 2011, pps. 502, 505)

Omitting the input of marginalized and vulnerable populations from early discussions of SRM leaves nobody available to spot any potentially damaging value judgments made in climate models. The result could be less critical science and less critical technology, potentially leading to the compounding of existing climate injustice.

The idea that vulnerable populations should be included at the very earliest stage of SRM research and development brings up a number of practical questions as well as questions about authority. It is not obvious how stakeholders from countries that may have no geoengineering research program can be included in what might be a highly technical discussion about modeling taking place in a lab located on a different continent. But with a little effort, there clearly are ways to engage those from the margins in the early days of SRM research and development.[10] Olson (2011) details a range of methods of upstream engagement in technological development, two of which lend themselves particularly to the participation of representatives from vulnerable populations. "Lab-scale Intervention" involves embedding outsiders directly in laboratories to raise questions in the physical location in which technological development decisions are being made. One could create international research teams that deliberately include specialists from vulnerable nations to help frame research questions, select important parameters, weigh the relative value of variables, establish initial conditions, and determine what to look for in results. Moving beyond the scientific community, Ethical, Legal, and Social Implications (ELSI) studies would introduce experts in law, ethics, and social science from the relevant countries to highlight particularly difficult challenges that the proposed technology will need to satisfy in their own countries. This type of embedding could be required for publicly funded global technologies such as SRM with team-members from vulnerable populations mandated by funding agencies.[11] The cost of such inclusion can be minimal, including, at most, some international air travel and more often simply a good Internet connection.

Participatory Technology Assessment (PTA) is a second way to encourage engagement in the process of technological development. PTA incorporates "citizen participation methods to complement expert analysis" (Olson, 2011, p. 33). Such participation could help evaluate whether "impacts" of proposed geoengineering schemes were being evaluated fairly from the perspective of those most likely to suffer them. Less rain in certain portions of Africa might matter considerably more than less rain in Wisconsin. PTA could also offer a fairer evaluation of whether the distribution of those impacts would match desert. To pursue these types of participation, it is not necessary to suppose all participants have equal cognitive authority in all of the technical details. It is necessary only to accord each participant the intellectual authority to participate in such upstream engagement (Longino, 2002). Such attempts at engagement could provide much needed opportunities for those from vulnerable countries to steer research and development of SRM technologies in directions that do not end up compounding the injustice that already exists in climate change.

CONCLUSION

The reasons for including the voices of vulnerable peoples in the earliest days of research into SRM technologies therefore turn out to be both substantive and normative. Substantively, it is quite possible that problematic assumptions common to a homogeneous community of research scientists can be corrected if there is adequate openness to criticism from those normally residing outside that community. Input gathered from vulnerable populations is particularly important for a technology with impacts as global and uncertain as SRM. Given the huge moral deficit that has accrued, some of the most important input will arguably come from those who will suffer the most harm from global warming, those who are least responsible for its occurrence in the first place, and those who have the potential to be most harmed through the uneven effects of SRM. The fact that these groups mostly overlap in sub-Saharan Africa, parts of Asia, and the Arctic makes it easier to identify those whose input into the development of SRM technologies is most important.

Furthermore, this inclusiveness is desirable, not just for the substantive reason that this will likely produce a better technology, but also for important normative reasons. On what grounds would the rich countries, having caused the problem in the first place and pursuing a solution that comes with clear attendant risks that will be imposed on others, want to exclude the input of those most vulnerable to harm? Having meaningful input into the development path of SRM technologies is, surely, the very least these populations deserve.[12]

NOTES

1. The dismantling of the annex/non-annex distinction at COP 17 in Durban in 2011 may prove an important step forward.

2. The language of moral deficit is designed to parallel the idea of "climate debt." This debt represents money rich countries are said to owe to the poor countries for taking up more than their fair share of the greenhouse gas absorbing capacities of the atmosphere (Bond, 2010).

3. Social science data on exactly how interested developing countries are in SRM is not yet available. Studies are underway at the University of Montana to address this question.

4. The extent of this vulnerability is still a matter of debate. Moreno-Cruz et al. (2011) find that inequalities in the effects of SRM may not be as great as some (e.g., Robock et al. 2008) have feared.

5. In the rest of this paper, I will use the term "climate injustice" to refer both to the wrongs that already result from anthropogenic climate change and to the wrongs that could potentially result from the SRM designed to treat it.

6. Sher's book addresses desert only as it applies to individual persons. Here I step outside his usage to talk, somewhat generally, about what whole countries do or do not deserve as matters of diachronic fairness.

7. Stirling's use of "normative" and "substantive" is directed at public engagement in general. He does not turn this argument particularly to vulnerable people.

8. What follows should therefore not be interpreted as an attempt to convincingly demonstrate the presence of non-epistemic values in global climate models. Such work belongs in a different literature. It is intended as a provocation to support the idea that, in the particular case of SRM, there is a potential value to including the views of those most vulnerable and most marginalized even at the earliest stages of the research processes. The argument for the value of such participation at later stages would then follow with considerably more ease.

9. Efforts to monetize all these costs based on willingness-to-pay models suffer from a range of notorious problems (Gowdy & Olsen, 1994).

10. Fuller discussions about how to include stakeholder participation can be found in the social science literature (Wilsdon & Willis, 2004; Leach, Scoones, & Wynne, 2005)

11. The US National Science Foundation is supporting an ELSI type of study on geoengineering that has resulted in this book (Grant number: SES # 0958095).

12. Thanks to Marion Hourdequin and Andrea Gammon for comments on a draft of this paper.

Chapter Six

Solar Radiation Management and Nonhuman Species

Ronald L. Sandler

As long as there has been climate, there has been climate change, and with it, ecological change. And as long as there have been systems of living organisms, adaptation to ecological change has occurred, and species populations have come into and gone out of existence as a result. However, the pace of ecological change is not fixed, and the greater the magnitude and rate of change, the more difficult is the challenge of adaptation for nonhuman (and human) populations. The distinctive feature of anthropogenic global climate change is highly accelerated levels of climatic and ecological change, in comparison to the recent historical past. As a result, extinction rates are expected to increase dramatically. Studies have found that 35% of bird species and 52% of amphibians have traits that put them at increased risk of extinction due to global climate change (Foden et al., 2008); that 20% of lizard species are likely to be extinct by 2080 due to global climate change (Sinervo et al., 2010); and that 15–37% of species will be committed to extinction by 2050 on mid-level warming scenarios (Thomas et al., 2004).

Nonhuman species populations that cannot quickly change their geographical ranges, such as those that disperse seed only locally or migrate slowly, are less able to adapt to rapidly changing ecological conditions than are those that are more mobile. For them, suitable habitat might contract, shift, or otherwise disappear more quickly than they are able to adjust.[1] Mountain and small island populations are also highly vulnerable, given the geographical limits (e.g., mountain tops and coasts) on their capacity to migrate as their environments change. So, too, are populations of species, such as corals, that are dependent upon very particular environmental conditions or on other species (Urban et al., 2012). Species whose members are more ecologically flexible with respect to diet and habitat, such as starlings and house mice, are less vulnerable. Moreover, populations of species that have fewer offspring and longer developmental periods (e.g., large mammals and hardwoods) are less likely to be able to biologically adapt to changing ecological conditions than are populations of species that reproduce rapidly and abundantly (e.g., weedy plants). Overall, the Intergovernmental Panel on Climate Change (IPCC) concludes:

> There is medium confidence that approximately 20-30% of species assessed so far are likely to be at increased risk of extinction if increases in global average warming exceed 1.5-2.5 C (relative to 1980-1999) [i.e., low warming scenarios]. As global average temperature increase exceeds about 3.5 C [i.e., mid-level warming scenarios], model projections suggest significant extinctions (40-70% of species assessed) around the globe. (IPCC, 2007a, p. 54)

Given that the background historical rate of extinctions is one species per million per year (or .0001%),[2] this constitutes a dramatic increase in extinctions rates, and "there are very strong indications that the current rate of species extinctions far exceeds anything in the fossil record" (Magurran & Dornelas, 2010, p. 3504).

Global climate change is an anthropogenic ecological disaster, and species extinctions are at the center of it. In this chapter, I discuss the implications of this for the ethics of geoengineering in general and solar radiation management (SRM) in particular. As the term is used in what follows, *geoengineering* refers to technology oriented projects intended to have large impacts on ecological and climatic processes in ways other than by reducing greenhouse gas emissions. *SRM* refers to geoengineering projects that involve intentionally modifying the amount of solar radiation that enters earth's atmosphere.

In the next section, I defend an account of the value of species according to which they have both instrumental and non-instrumental (or final) value. In section following, I develop an argument in favor of geoengineering on the basis of the value of species. Finally, I evaluate the argument in support of geoengineering and assess the extent to which it favors SRM in particular. The overall conclusion of the discussion is that the value of species supports aggressive action to keep ecological and evolutionary relationships intact but that SRM would largely fail to do so. Therefore, the value of species is not supportive of SRM.

THE VALUE OF SPECIES AND WHY IT IS WRONG TO CAUSE THEIR EXTINCTION

Species are historical phenomena. They come into existence at a particular time and place, they change and adapt over time, and they eventually go extinct. The vast majority of species that have ever been on earth already no longer exist. If the fate of all species is extinction, then why should the fact that anthropogenic climate change will result in an accelerated rate of extinction be problematic? At most, greenhouse gas emissions resulting from human activities are a contributing cause to hastening something that would have occurred anyway. Moreover, there have been several mass extinctions in the past. Why should the fact that they were the product of entirely natural events (far predating the evolution of *Homo sapiens*), whereas current climatic change is partly forced by human activity, make any ethical difference?

There are two primary difficulties with the reasoning described above. The first is that the fact that something occurs in nature or by natural processes does not imply that it is permissible for humans to do or bring about (Mill, 1904). States of affairs and events that are abominable when intentionally done by humans are ubiquitous in the natural world—forced copulation, territorial killing, and selective reduction of newborns, for example. What happens in nature is not, in virtue of its occurring there, permissible for us to do or cause. The second problem is that sometimes it is unethical to "merely" hasten the occurrence of something that would have eventually happened anyway. All people will eventually die. In fact, the vast majority of people who have ever lived are no longer alive. But it does not follow from this that it is permissible to engage in activities that will hasten their deaths, even if the contribution is unintended (e.g., negligence) rather than intentional (e.g., murder).

The reason that it is wrong to hasten the death of people is that people are a locus of value. They matter for what they are. Therefore, we need to take them into account in practices, policies, and actions that affect them. Do species likewise have value that we need to care about and that makes it wrong to prematurely end their existence? This is a perennial issue in environmental ethics.

It is widely recognized that individual species and biodiversity generally have instrumental value. Instrumental value is use value. Something has instrumental value just in case it is an effective means to a desired or worthwhile end. Individual species are instrumentally valuable in myriad ways. Some species are medicinally valuable (e.g., horseshoe crabs and sweet wormwood), some are economically valuable (e.g., honey bees and mahogany), and some are recreationally valuable (e.g., rainbow trout and sequoia redwoods). Not all species are equally instrumentally valuable. In fact, the vast majority of species are not keystone species or economically or medically significant. But all species have at least some instrumental value—e.g., scientific value and option value (the possibility of being valuable in the future) (Hunter, 2001). Biodiversity is also instrumentally valuable in myriad ways. For example, plant species richness has been found to enhance ecosystem multifunctionality (Maestra et al., 2012); restoration of biodiversity has been found to increase ecosystem services and productivity (Benayas et al., 2009; Worm et al., 2006); and maintaining biodiversity appears to be associated with a lower prevalence of infectious disease transmission (Keesing, 2010).

Particular species and biodiversity also have enrichment value. Individual species can be beautiful, bizarre, and amazing, and the diversity of the biological world is astounding. Because of this, nonhuman species make the world a more interesting place. They enrich our experience. Moreover, it is beneficial for a person to cultivate appreciation and wonder toward the variety and complexity of biological diversity as well as toward the beauty and uniqueness of individual forms of life. Such attitudes open her to rewarding and enjoyable relationships and experiences (Sandler, 2007; O'Neill, 1993; Cafaro, 2001a; Sarkar, 2005; Carson, 1956). Thus, one reason it is wrong to cause species to go extinct, particularly on a massive scale, is that it is bad for us. Biodiversity contributes to ecosystem services that we depend upon, and individual species provide all sorts of goods and experiences, from material to aesthetic, that we use and enjoy.

However, instrumental value is not the type of value that primarily explains why prematurely ending the life of a person is problematic. After all, killing a person is wrong even if she is not particularly helpful. The reason for this is that people possess final value. They have value in themselves or for what they are.[3] Final value is importantly different from instrumental value in that it is not substitutable or replaceable and because it is not conditional on the final value of anything else. Because instrumental value is use value, it is always derivative on the value of the ends to which it is a means. A thing's instrumental value fluctuates based on changes in the value of the end, the circumstances that obtain, and whether alternative or more efficient means are available. For example, a fishing line has instrumental value just in case a person wants to catch fish, and its value might diminish if she gains access to a much more effective fishing net or moves to a location where fishing opportunities are limited. Moreover, if one fishing line is lost but many equally good replacements are available, there is no net loss in instrumental value. Final value is not like this. Because a person has final value, her value is not derivative upon the value of anything else, and her value is not altered by more people coming into existence. If she dies, there is a value loss, even if another (equally capable) person is born. It is because people have final value, not just instrumental value, that they are not substitutable and replaceable.[4]

It is a common view among environmental ethicists that species also have final value. Holmes Rolston III, a prominent and influential environmental ethicist, has argued that each species is a locus of final value because each is a distinctive historical form of life that is the product, process, and instrument of creative and generative evolutionary processes. On his view, a species possesses final value because it is a unique and potentially productive evolutionary trajectory (Rolston, 1995, 2001). Species are to be protected because a species extinction "shuts down the generative processes, a kind of superkilling. . . . To kill a particular plant is to stop a life of a few years, while other lives of such kind continue unabated, and the possibilities for the future are unaffected; to superkill a particular species is to shut down a story of many millennia and leave no future possibilities" (Rolston, 1995, p. 523). Thus, on Rolston's view, species have final value in virtue of their unique ecological and evolutionary situatedness. His is a natural historical account of the final value of species, which are prevalent in environmental ethics (Preston, 2008; Katz, 2000; Cafaro, 2010b; Soulé, 1985).

On Rolston's view, the final value of species and biodiversity is objective, in the sense that they possess it in and of themselves, independently of whether anyone values them. As he puts it: "These things count, whether or not there is anybody to do the counting" (Rolston, 1982, p. 146). Other environmental ethicists have argued that species possess final value but that they do so because we value them for what they are or that they are and not merely for what they do for us. As Baird Callicott, a prominent proponent of this view, has put it: "[Species] may not be valuable *in* themselves but they certainly may be valued *for* themselves. According to this . . . account, value is, to be sure, humanly conferred, but not necessarily homocentric" (Callicott, 1989, p. 151).[5] The claim that people value species as ends or for what they are, and not merely instrumentally, is not speculative. It is supported by empirical research and evidenced by the very large memberships of environmental and species conservation organizations (Bosso, 2005; Kempton et al., 1995).

Moreover, the properties in virtue of which people value species and biodiversity are much the same as those in virtue of which Rolston and others believe that they have objective final value—i.e., their human independent ecological and evolutionary situatedness. What people value is species expressing their distinctive form of life in their evolved ecological context— migrating wildebeest, soaring condors, spawning salmon, towering torreya, dancing honey bees, and breaching humpbacks. When their naturalness and wildness is taken away, when they are removed from their habitats, species do not have the same value. It is polar bears in the Arctic, not polar bears in the Central Park Zoo, that are able to fully pursue the polar bear form of life, which is what people find magnificent about them. Again, this is not speculative, it is evidenced by research on people's environmental values, the resources that people contribute to wilderness protection and *in situ* species conservation, support for legislation (such as the U.S. Endangered Species Act) that is committed to preserving species in their habitat, and the priority placed on *in situ* preservation in conservation biology.

Whether the final value of species is only valuer dependent, or if it is valuer independent as well, is an interesting and important issue in environmental ethics. However, it is not one that needs to be settled here. What is crucial for present purposes is that species possess final value. Moreover, they do so in virtue of their ecological and evolutionary relationships as well as their independence from human impacts and control.[6] An implication of this is that they retain their final value only so long as those relationships remain intact and are not disrupted by human activities. (The ecological value of species and biodiversity—a prominent type of instrumental value mentioned earlier—is similarly contingent on maintaining ecological relationships.) As a result, "It is not preservation of *species* that we wish, but the preservation of *species in the system*. It is not merely *what* they are, but *where* they are that humans must value correctly. . . . The species can only be preserved *in situ*; the species *ought* to be preserved *in situ*" (Rolston, 2001, p. 411).

AN ARGUMENT FOR GEOENGINEERING

Anthropogenic global climate change will severely disrupt the ecological relationships of a large proportion of species, resulting in highly elevated rates of species extinctions and so a large amount of value loss. Because the final value of species is tied to their ecological and evolutionary situatedness, the value of those species cannot be fully preserved through *ex situ* conservation—e.g., by maintaining populations in zoos, botanical gardens, or seed banks. Nor can it be preserved by translocating species beyond their historical range and establishing new independent wild populations, as some conservation biologists and environmental ethicists have proposed, because that also would not maintain their historical ecological relationships (Hoegh-Guldberg et al., 2008; U.S. Climate Change Science Program, 2008; Vitt et al., 2010; Richardson et al., 2009; Minteer & Collins, 2010; Camacho et al., 2010; Thomas, 2011).[7] The only way to preserve the full value of species is to prevent (or minimize) disruption of their ecological relationships from occurring. Therefore, the final value of species strongly favors prioritizing reducing the anthropogenic forcing of the climate system (and so ecological systems) rather than managing and adapting to ecological change.

A large amount of anthropogenic climate change is already "locked-in" in virtue of greenhouse gas emissions that have already occurred. Even a complete cessation of future emissions will not prevent significant ecological changes from occurring (Gillett et al., 2011). One reason for this is that many greenhouse gases take a considerable amount of time to break down in the atmosphere. According to the IPCC (2007), "[A]bout 50% of a CO_2 increase will be removed from the atmosphere within 30 years and a further 30% will be removed within a few centuries. The remaining 20% may stay in the atmosphere for many thousands of years." Therefore, the elevated atmospheric concentrations of greenhouse gases that are causing the current anthropogenic climate forcing are the product of the cumulative emissions since the industrial revolution. Moreover, even if global emission levels were to flatten out, there would be a higher concentration of CO_2 in the atmosphere in fifty years than there is now, because CO_2 emission rates would still be greater than the rate at which CO_2 is removed from the atmosphere. Another reason that so much climate change is "locked-in" is that the effects of elevated atmospheric greenhouse gases levels can take a considerable amount of time to manifest in climatic and ecological systems—e.g., effects on precipitation patterns, plant growth, ocean acidification, and mean surface air temperatures. Moreover, once the effects begin to occur, they can be persistent and cascading, particularly if there are significant climatic tipping points and feedbacks—for example, if ocean currents are significantly altered, if ice sheets rapidly break apart, or if large amounts of methane previously trapped under permafrost is released.

Because the final value of species favors minimizing anthropogenic ecological change over adaptation to ecological change, it supports practices and policies that would reduce and arrest the climatic momentum described above. There are, in general, two ways to do this. The first is to mitigate future emissions aggressively. Although a significant amount of anthropogenic climatic change is "locked-in," the magnitude of the change varies dramatically according to different future emissions scenarios. As figure 6.1 illustrates, the range in global mean surface air temperature increases from the bottom end of the low future emissions path to the upper end of the business-as-usual emissions path is 2.1–14.3 degrees Fahrenheit, and the difference in atmospheric concentrations of CO_2 and CO_2e (carbon dioxide equivalent[8]) between the two is 520 ppm and 945 ppm respectively (Climate Interactive, 2011). (Current atmospheric CO_2 concentrations are approximately 394 ppm.) As discussed above, the adaptation challenge associated with global climate change is a product of the rapidity and magnitude of the change. The greater the change, the more ecological relationships are disrupted, the more species go extinct, and the greater the value loss (both final and instrumental). Therefore, the value of species strongly favors policies and practices that would accomplish a lower emissions path.

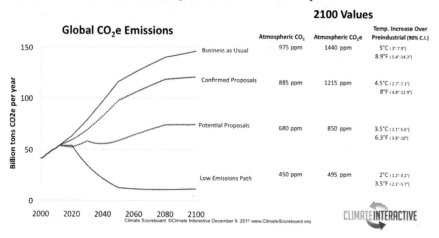

Figure 6.1.

The primary source of increases in atmospheric CO_2 (and overall greenhouse gas) levels is the combustion of fossil fuels—e.g., oil, coal, and natural gas (IPCCa, 2007). Therefore, the value of species favors reducing fossil fuel consumption. This could be accomplished through increases in efficiency (e.g., building insulation, improved vehicle mileage per gallon, smart grids, and compact florescent light bulbs), changes in lifestyle (e.g., less meat consumption, fewer air flights, smaller families, and lower thermostat settings in the winter), and developing alternative energy sources and infrastructures (e.g., solar, wind, and geothermal).[9]

Another way in which atmospheric greenhouse gas levels could be reduced from the business-as-usual scenario is by sequestering potential greenhouse gases. For example, carbon is sequestered in biomass, so future greenhouse gas concentrations can be reduced by protecting biomass rich areas, such as forests.[10] Carbon is also sequestered in the soil, and this can be increased by widespread use of agricultural practices that promote rather than diminish the carbon richness of topsoil. Some forms of geoengineering also take a sequestration approach to reducing atmospheric CO_2 concentrations. For example, carbon capture and storage (CCS) involves capturing CO_2 at its source point—fossil fuel power plants—and pumping it in geological formations or aquifers where it would be trapped (either in gaseous form or by catalyzing a reaction to turn it into carbonate). And ocean fertilization involves seeding the ocean with iron (or other fertilizers) to increase phytoplankton growth and thereby carbon absorption, which is then sequestered when the phytoplankton die and descend to the ocean floor. Some atmospheric scrubbing technologies, including filtering CO_2 out of ambient air, may also need to be paired with a storage or sequestration strategy.[11] (Other forms of atmospheric scrubbing—e.g., using catalysts to increase the rate of CO_2 degradation in the atmosphere—may not require a companion sequestration technology.)

The second way (i.e., in addition to reducing atmospheric greenhouse gas levels) in which the momentum of anthropogenic climate change could be arrested is by intervening in the climate system to forestall climatic and ecological changes that increased atmospheric greenhouse gas levels would otherwise cause. SRM is the most prominent candidate for how this might be accomplished. SRM does not address the amount of CO_2 in the atmosphere. It aims to reduce the climatic and ecological impacts of increased greenhouse gas levels by limiting the amount of solar radiation that enters the atmosphere. Strategies for SRM include stratospheric sulfur injections, space-based mirrors, and brightening marine clouds with seawater mist. If SRM is employed and is successful, then the increases in global mean surface air temperature predicted to be associated with elevated atmospheric concentrations of greenhouse gases would be mitigated or eliminated. As a result, SRM would reduce the magnitude of the ecological changes associated with increased atmospheric concentrations of greenhouse gases. This, in turn, would reduce species losses associated with global climate change in comparison to the same scenario without deployment of SRM technologies.

Here, then, is a summary of the argument in favor of geoengineering, based on the value of species:

1. Many species, and biodiversity, generally have instrumental value to people.
2. Many species, and biodiversity generally, have final value (or value in themselves) in virtue of their ecological and evolutionary situatedness as well as their independence from human control and impacts.
3. The final value of species and biodiversity is in principle not substitutable or replaceable, and the instrumental value of species and biodiversity is in practice often difficult to substitute or replace.
4. The greater the rate and magnitude of anthropogenic climate change, the more anthropogenic disruption of ecological relationships that will occur and the more nonhuman populations and species that will go extinct.
5. Therefore, the greater the rate and magnitude of anthropogenic climate change driven ecological change, the greater the nonhuman species instrumental and final value loses. (From 1 to 4)
6. Losses of value ought to be avoided.
7. Therefore, the value of species favors reducing the amount of ecological change associated with global climate change. (From 5 and 6)

8. Mitigation of future greenhouse gas emissions would reduce, but not prevent, significant ecological change from occurring because, even on low future emissions trajectories, a significant amount of global climate change is already "locked in" due to prior emissions.[12]
9. There are significant political, technological, structural, and attitudinal/behavioral impediments to accomplishing a low atmospheric greenhouse gas concentration trajectory through emissions reductions and traditional sequestration techniques (e.g., forest and soil sequestration) (Gardiner, 2006b).
10. Therefore, mitigating future greenhouse gas emissions and employing traditional carbon sequestration techniques are inadequate to prevent global climate change from resulting in significant losses of nonhuman species value. (From 8 and 9)
11. Geoengineering can reduce atmospheric greenhouse gas levels as well as mitigate the amount of ecological change associated with elevated atmospheric greenhouse gas levels beyond what is possible through mitigating future emissions and non-geoengineering approaches to sequestration.
12. Therefore, the value of species favors pursuing geoengineering approaches to reducing the ecological impacts of anthropogenic climate change. (From 7, 10, and 11)

I believe that this argument constitutes the strongest pro-geoengineering case that can be made based on the value of species. In the next section I evaluate just how strong it is both in general and with respect to SRM in particular.

EVALUATING THE CASE FOR SRM

The argument in support of geoengineering developed in the previous section is a *ceteris paribus* (other things being equal) argument because it concerns only the value of species. There are many other values and ethical considerations relevant to whether, how, and under what conditions it is justifiable to engage in geoengineering. Geoengineering raises issues related to governance, human rights, justice, future generations, economic costs, and human well-being. An all-things-considered conclusion would require taking all of those considerations into account as well. Therefore, the argument, even if sound, only justifies the conclusion that one particular consideration—i.e., the value of nonhuman species—favors engaging in geoengineering.

The argument is also general. It does not disambiguate different forms of geoengineering. However, different forms of geoengineering have different capacities for reducing ecological disruptions, and different approaches to geoengineering have quite different ethical profiles—e.g., distributions of burdens and benefits, regulatory frameworks and oversight mechanisms, ancillary effects, and risk profiles. To determine the extent to which the argument supports a particular form of geoengineering, it needs to be particularized to the type of technological intervention as well as contextualized to the conditions of the intervention.

Moreover, the argument is predicated on the premise that developing and implementing geoengineering would successfully reduce the disruption of historical ecological relationships, and thus the adaptation challenges faced by species populations. This premise involves several assumptions: (1) geoengineering would effectively accomplish its intended technical aim—e.g., sequestering billions of tons of carbon in saltwater aquifers or reducing the amount of solar radiation that enters the atmosphere; (2) there would not be significant unintended effects associated with deployment of the technology that would disrupt ecological relationships; (3) accomplishing the intended technical aim would reduce the climatic and ecological changes associated with global climate change; and (4) pursuing and implementing geoengineering would not displace other, possibly more justified, approaches to reducing ecological disruptions associated with global climate change.

With respect to the first and second of these assumptions, it is crucial to note that the context of geoengineering deployment is one of tremendous climatic and ecological uncertainty. The large amount of uncertainty is evidenced by the very wide range of possible global mean surface air temperature increases between and within different future emissions scenarios. Intervening effectively into highly complex ecological systems, particularly longitudinally, is difficult under any circumstances. However, it is particularly difficult when some elements of the systems are not well understood (e.g., tipping points and climate feedbacks), variables are indeterminate (e.g., future emissions), and background conditions are rapidly changing. Each of these exacerbates the difficulty with determining what the "right" amount of engineering would be—e.g., how much solar radiation to try to prevent from entering the atmosphere or how much carbon to sequester.

Moreover, identifying and controlling the impacts of large technological interventions in ecological systems has not been done well in the past, even on local or regional scales and even absent the challenge of rapid climatic change. For example, intensive, technologically oriented industrial agriculture, which has transformed terrestrial ecosystems as much or more than any other activity, has resulted in diminished topsoil, reduced fresh water availability, aquatic dead zones, biodiversity losses, and large amounts of greenhouse gas emissions. Industrial agriculture has, for the most part, been successful in accomplishing its goal: increasing yields. But it has done so with expensive inputs and enormous externalized costs, both ecological and social.

Geoengineering involves intervening in atmospheric, climatic, and ecological systems at a global scale using technologies that cannot be tested on comparable scales within like systems in the context of highly contingent and uncertain variables (Robock et al., 2010).[13] Given this, and in light of the outcomes of prior large scale interventions, it is reasonable to believe that unanticipated and unwanted detrimental side effects are likely to occur and be difficult to address. For example, concerns have been raised that SRM might alter precipitation patterns, which could have transformative impacts on local ecosystems, and, by diminishing radiation reaching the earth's surface, SRM might reduce biological productivity and, thereby, food production.

In a context marked by amplified climatic and ecological uncertainty, technologies that are more control oriented and precision dependent are likely to be less successful than those that are not, and technologies that are more interventionist into complex and inadequately understood systems are likely to have greater unanticipated effects than those that are not. This is a function of complexity and uncertainty in dynamic and integrated systems. The greater the complexity, unpredictability, possibilities, and uncertainties, the more circumspect we should be regarding our capacity to intervene effectively and responsibly, particularly at a global scale. These concerns apply quite strongly to SRM as SRM involves direct intervention into climatic systems to moderate an aspect of those systems: solar radiation, which is crucial to many critical ecological and climatic processes. Not all forms of geoengineering have these features. For example, CCS does not involve directly manipulating climatic and ecological systems; the intervention is more localized than is the case with SRM, and CCS's success does not depend upon "tuning" ecological or climatic systems to the extent that SRM's does.[14]

The third assumption—i.e., that geoengineering would address the significant climatic and ecological dimensions of increased atmospheric greenhouse gas concentrations—is not satisfied by many geoengineering proposals. For example, even if SRM successfully reduced the global mean surface air temperature by preventing some solar radiation from entering the atmosphere, it would not address the serious problems associated with carbon dioxide accumulation and ocean acidification (Robock et al., 2009; Hegerl & Solomon, 2009; Wigley, 2006). Moreover, SRM raises considerable termination problems as a result. So long as greenhouse gases continue to accumulate in the atmosphere, the need to maintain solar radiation management continues. SRM does not address the underlying cause of the potential for ecological disruptions but rather prevents some of the disruptions from occurring. As long as the potential for disruption remains (i.e., elevated levels of greenhouse gases) SRM will be required.

Assessment of the fourth assumption—i.e., that pursuing and implementing geoengineering would not displace other, more preferable approaches to reducing ecological disruptions associated with global climate change—is of necessity somewhat speculative. It is difficult to know how pursuing one response to global climate change will affect another. Aggressively mitigating greenhouse gas emissions is likely to have non-trivial economic and individual costs. Therefore, if geoengineering is presented as a comparatively inexpensive climate change "solution" that does not require lifestyle or cultural changes, then it may function as a disincentive to individual behavioral changes as well as an additional barrier to the already difficult challenge of climate change policy enactment. However, geoengineering in general and SRM in particular typically have not been presented as climate change solutions. Proponents of SRM defend its development as a possibly necessary "last resort" to avoid catastrophic climate change, one that should be used only if emissions mitigation and traditional sequestration efforts prove inadequate (Crutzen, 2006).[15] It is possible that SRM will still function as a disincentive to mitigation when presented in this way, but it is less likely to do so than if it were presented as an alternative climate change solution. However, to the extent that this concern has merit, it applies particularly strongly to SRM. The reason for this is that SRM, even if successful, would not address many of the problematic aspects of greenhouse gas accumulation, such as ocean acidification. So it is not an alternative to emissions mitigation (and sequestration) for reducing many of the types of ecological disruptions associated with global climate change. Presenting it or perceiving it as such would, therefore, involve seriously problematic misrepresentation.[16]

Some forms of geoengineering may have considerable promise for forestalling significant ecological changes and associated nonhuman population extinctions and, thereby, for preserving the value of some species. However, SRM is not among them. Successful SRM, unlike some other forms of geoengineering, requires large-scale and direct intervention into crucial aspects of climatic processes. There are strong considerations, based on past technological interventions into complex ecological systems and amplified by the distinctive features of global climate change, that suggest that such interventions are likely to be difficult to control as well as likely to result in unintended and significantly ecologically problematic effects. Moreover, SRM, because it involves large-scale direct intervention, would, even if successful, alter the ecological context of many species to at least some extent—e.g., through changes in radiation (light) patterns. SRM does not eliminate or reduce anthropogenic greenhouse gas accumulations. It aims to prevent them from affecting global mean surface air temperature through further human intervention. As a result, it would not be an effective approach to preserving the human independent natural historical value of species. Furthermore, SRM, even if successful, would not address several of the ecologically problematic impacts of increased atmospheric greenhouse gas concentrations, such as ocean acidification. Finally, promoting and pursuing SRM would be highly problematic, if doing so were to provide a disincentive to emission mitigation and non-geoengineering sequestration approaches to reducing the magnitude of ecological change associated with anthropogenic greenhouse gas emissions.

CONCLUSION

The value of species strongly favors responses to global climate change that would help forestall population extinctions by maintaining historical ecological relationships. This includes aggressive mitigation of future emissions. It may also include some forms of technology for assisted sequestration, such as CCS.[17] However, the value of species does not support SRM. SRM involves intentional manipulation of climatic systems; it does not address many of the ecological effects associated with increased atmospheric greenhouse gas concentrations, and it is likely to be difficult to predict and control. Because SRM is unlikely to be successful in preserving significant numbers of species and, even when it is successful, the properties of species in virtue of which they are valuable are likely to be undermined, it is not supported by the value of species. This is not, however, to conclude that the value of species is opposed to SRM. SRM may not be an effective way to improve the situation of nonhuman species populations under conditions of anthropogenic global climate change. But, particularly on the higher emissions scenarios, climate change is expected to be so disruptive of the ecological relationships of such a large proportion of species that it is doubtful that SRM will make things worse for them. That is, there may be contexts that would already be sufficiently bleak from the perspective of the value of nonhuman species that SRM would not pose a significant further threat to their value.

NOTES

1. It is estimated that species ranges are shifting altitude at a median rate of 11.0 meters per decade and in latitude at a median rate of 16.9 kilometers per decade in response to global climate change (Chen et al., 2011). See also Burrows et al. (2011).

2. Baillie et al. (2004) calculate the historical rate of extinction as 0.1-1 E/MSY.

3. It is common in environmental ethics to refer to non-instrumental value as "intrinsic value." However, the "intrinsic value" terminology can be ambiguous and misleading. In addition to being used to refer to non-instrumental value, "intrinsic value" is also often used to mean "objective value" and "nonrelational value." But these are conceptually distinct. It is possible that there be non-instrumental value that is subjective rather than objective in the sense of being dependent upon the evaluative attitudes of valuers. It is also possible that there be non-

instrumental value that is based on relational rather than non-relational properties. Because it is possible to have non-instrumental value that is dependent upon evaluative attitudes and/or based on relational properties, it is misleading to refer to non-instrumental value as "intrinsic." In contrast, "final value" captures what is crucial about non-instrumental value—i.e., it is value that is not for the sake of anything further—without the ambiguities or misleadingness associated with "intrinsic value."

4. That each person is a distinctive locus of value is why human rights violations are not substitutable, for example. For a discussion of the implications of this for the ethics of climate change, see Caney (2008, 2010).

5. See also Elliot (1992).

6. There are exceptions to this. Some species may be valued for what they are for reasons that are not dependent upon their ecological properties—e.g., some culturally or religiously significant species.

7. For extended discussion of the ethics of assisted colonization, see Sandler (in press a).

8. Carbon dioxide equivalent refers to the concentration of CO_2 that would have the equivalent warming potential as the concentration of all greenhouse gases taken together.

9. For a discussion of the ethics of mitigation, including criteria for evaluation different emissions reduction strategies, see Sandler (in press b).

10. This is the motivation behind the United Nations Framework Convention on Climate Change's (UNFCCC) Reducing Emissions from Deforestation and Forest Degradation (REDD) program.

11. Ideally, CO_2 captured from the point of combustion or filtered from ambient air would not merely be stored but used. Algae that "feeds" on CO_2 and efficiently produces fuel (or fuel precursors) in bioreactors (or other industrial-scale forms) would be an instance of this.

12. As discussed earlier, even on lower emissions trajectories, the IPCC predicts that 20 to 30% of species will be at increased risk of extinction.

13. Small-scale testing and pilots are possible. For example, feasibility testing for a cloud brightening approach to SRM is scheduled for 2012, and pilot CCS projects are underway (Fecht 2011a, 2011b).

14. This is not to endorse CCS, because it is not an all-things-considered evaluation of its feasibility, ecological impacts and risks, or broader ethical profile.

15. Proponents of CCS typically defend it as a "bridge technology" rather than a long-term climate change solution. The idea is that it will help to reduce atmospheric greenhouse gas levels until a less carbon intensive infrastructure and economy is established (Chu, 2009).

16. Gardiner (2010) has argued that "arming the future" with SRM is in itself ethically problematic because it would place future generations in a position where they have to make a tragic choice between engaging in massive geoengineering or allowing runaway climate change. Because we would be responsible both for not mitigating to prevent climate change and for developing the SRM technologies, it would be a situation of our making. From the perspective of nonhuman species, the situation would be tragic because both alternatives, SRM and runaway climate change, fail to maintain the ecological relationships on which species populations and their value depend.

17. Ocean fertilization sequestration would not be supported because it, like SRM, involves further direct intervention and manipulation of ecological systems.

Moral Hazards and Hidden Benefits

Chapter Seven

The World That Would Have Been

Moral Hazard Arguments Against Geoengineering

Ben Hale

According to the United Kingdom's Royal Society Report Geoengineering the Climate, "one of the main ethical objections to geoengineering" is the purported "moral hazard" (Royal Society, 2009, p. 39). Roughly speaking, the moral hazard is the complication that the successful deployment of a geoengineering technology or, in some cases, the mere possibility of the successful deployment of a geoengineering technology, may cause individual or collective actors to turn attention away from alternate solutions to the climate crisis. A cursory review of the arguments, however, suggests that there is a good deal of confusion about what, exactly, the unique moral hazard associated with geoengineering entails.

For instance, David Keith characterizes the moral hazard as a concern that geoengineering will weaken a commitment to cutting emissions. He says: "knowledge that geo-engineering is possible makes the climate impacts look less fearsome. And that makes a weaker commitment to cutting emissions today. This is what economists call a moral hazard" (TED talk, September 2007).[1] Martin Bunzl, by contrast, characterizes the moral hazard slightly differently, claiming that geoengineering may in fact stimulate an increase in carbon output (Bunzl, 2008). Where for Keith it would appear that the problem lies in permitting business as usual, for Bunzl it would appear that the problem lies in encouraging bad behavior. The aforementioned Royal Society Report offers yet a third interpretation, suggesting not only that geoengineering will reduce support for mitigation policies but also that it will "divert resources from adaptation" (Royal Society, 2009, p. 4). For the Royal Society Report, it would appear that the moral hazard lies in the diversion of resources. Clearly, the problems are related but quite distinct in their emphasis.

Is the problem that geoengineering may inspire fewer people to take notice of the globalized impacts of their actions? That people will refuse to change their behavior if a geoengineering technology is deployed? That people *will* change their behavior if a geoengineering technology is deployed? There is little agreement amongst geoengineering discussants about what exactly the moral problem is. If the moral hazard argument against geoengineering is to carry any persuasive weight, then it must clearly specify what changes in behavior will follow from geoengineering. Unfortunately, articulations of the moral hazard are too ambiguous to do this work. More vexingly, however, there are many potential hazards associated with the moral hazard—environmental, psychological, sociological, political, policy-related, and militaristic—any one of which would require a different response from policymakers. In this respect, the moral hazard is not only ambiguous but also vague. Without a clear articulation of the problem, moral hazard arguments may be sowing more confusion than adding clarification.

The problem with allowing this ambiguity and vagueness to stand is that the moral hazard has become a sort of catch-all used to refer to a suite of objections and hazards. Put somewhat illustratively, if a new technology emerges that will enable humans to eat as much as they want without putting on weight or suffering any health effects, some may object that the technology introduces a moral hazard. It may well do so. People will be inclined to change their dietary habits in the face of such a technology, and there may well be moral problems with such changes in behavior. Perhaps they will begin eating more gluttonously, which is bad for their character, or perhaps for the environment, or perhaps for animals. Perhaps, instead, there are better ways for them to live rewarding lives without the use of such a technology, in which case the problem is with what they will miss out on. On the other hand, perhaps their lives will be made infinitely richer, for now they will be empowered with the possibility of eating many delicacies that were otherwise too difficult for them. The moral hazard describes all of these cases, among many others, and when used without further specification, invites rebuttals that often prove more confusing than helpful.

This, in fact, is precisely how such arguments are playing out. As currently used, the alleged "moral hazard" associated with geoengineering functions as a falsely concrete straw man and, therefore, is both easy to offer as a criticism and equally easy to dismiss. For instance, the 2009 Royal Society Report claimed that "concerns have been expressed that geoengineering proposals could reduce the fragile political and public support for mitigation and divert resources from adaptation" (Royal Society, 2009, p. 4). To address these concerns, the authors suggested that more research be done. Not a year after the release of the report, the Natural Environment Research Council (NERC) conducted just such a study. In their report, they found that a majority of those polled believed that "it would be ethically and practically important to link any new climate change solutions to continued mitigation" (Godfray, 2010, p. 1). From this, NERC concluded that "this evidence is contrary to the 'moral hazard' argument that geoengineering would undermine popular support for mitigation or adaptation" (Godfray, 2010, p. 2). Somehow NERC was able to draw such a sweeping conclusion even though the mere fact that many people *believe* that geoengineering should also be accompanied by mitigation or adaptation does not speak to how people *will actually behave* in the face of a geoengineering regime. Only a small number of moral hazard arguments are concerned that the population holds the right attitudes or has the right beliefs (see, for instance, the Hubris Objection and the Attitude Objection below). The moral hazard is more often a concern about increased exposure to risk: about actual behaviors, not individual beliefs and/or anticipated behaviors following from these beliefs. The problem with the NERC study of moral hazards and geoengineering, however, goes well beyond a simple misapprehension of the moral hazard. It is a problem with the ambiguity and vagueness of the moral hazard argument against geoengineering altogether.

In this chapter, I argue that moral hazard arguments against geoengineering fail on their face. They fail not because they are wrong or incorrect but because they are far too complicated and multilayered to do the work that they are assumed to do. They are, as I argue here, both ambiguous and vague. Though moral hazard arguments ostensibly offer a compelling reason for avoiding geoengineering altogether, or at least proceeding cautiously, whatever moral hazards we identify do not present a *prima facie* reason—I am using the term in its Rossian sense[2] —for altering our approach to geoengineering. It is my view that geoengineering-related moral hazards are better addressed more directly with other arguments.

Building on previous work, I proceed first by offering a working definition of the moral hazard and of geoengineering. I then cover the variety of definitions of the moral hazard and clarify that moral hazards always require a supplementary argument to clarify what the wrong consists in. Second, I offer three variations on the generalized moral hazard claim. Each variation points to an important weakness with the successful deployment of geoengineering, but due to ambiguity, requires further careful articulation. Third, I introduce a compendium of hazards that might be said to accompany any of the three variant interpretations of the moral hazard. Coupled with the ambiguity of variations, the vagueness associated with the hazards proves crippling for the moral hazard as a stand-alone argument against geoengineering. Each of the variant hazards that falls under the penumbra of the moral hazard is a real concern, I believe, and each, in its way, stresses some increased risk associated with behavioral change following from geoengineering. Inasmuch as the hazards are separate from the so-called moral hazard, each must be assessed on its own terms. Hopefully this compendium of arguments and objections will prove helpful when sorting through the various articulations of moral hazard arguments.

MORAL HAZARDS AND MORAL THEORIES[3]

The moral hazard is a market failure most commonly associated with insurance but also associated by extension with a wide variety of public policy scenarios, from geoengineering, to corporate bailouts, to health insurance, to environmental disaster relief.[4] Loosely defined, a moral hazard is said to explain the occurrence of behavioral change in the face of insurance. More colloquially, it is sometimes considered to be "taking advantage of" insurance. For the purposes of this paper, it should suffice to define the moral hazard as "the danger that, in the face of insurance, an agent will increase her exposure to risk."

For instance, suppose that Smith insures his house against fire. The reason that he might insure his house against fire is so that he will be reimbursed if a fire breaks out. But certainly, if Smith is insured, he has less to worry about, and consequently, he has less incentive to be attentive to the devices that a more cautious homeowner might use to protect herself. In this case, as Kenneth Arrow notes, the "probability of fire is somewhat influenced by carelessness, and of course arson is a possibility, if an extreme one" (Arrow, 1963, p. 961). The presumed problem, of course, is a natural one: what use is insurance if the insured party changes his exposure to risk once he becomes insured?

Talk of moral hazards has been around since at least as long as the modern insurance industry, which some date back as far as 1662 (Hacking, 1975, 2003, p. 28). It was not until 1963, however, that Kenneth Arrow employed the concept to discuss the economics of medical care. By 1968, he and Mark Pauly had engaged in an exchange that was to invigorate the use of the concept in public policy economics for years to come. More than earlier investigations of the moral hazard, this exchange identifies the tendency of insurance coverage to change the behavior of individual actors (Arrow, 1963, 1965, 1968, 1985, Pauly, 1968).

Though the moral hazard is, in principle, easy enough to grasp, even within the economics community few are clear on the meaning of the term. There are unwritten ambiguities in Arrow's early formulation of the moral hazard mentioned above, for instance. On one hand, it appears that insurance forces a change in Smith's behavior such that he becomes more *careless*. On the other hand, it appears that insurance places Smith in a position in which he has greater incentive to take a deliberate action and torch his house to its foundations. It is not clear whether insurance encourages Smith to act or omit, for instance. This problem, among others, is endemic throughout the moral hazard literature. Bryan Dowd tries to salvage the idea of the moral hazard by explaining that there are in fact several manifestations of a moral hazard. "Insured individuals may exercise less caution (e.g., smoking) . . . insured individuals may seek professional intervention at a lower level of illness severity, demand a higher quality of care, or not shop as carefully for least cost providers as uninsured individuals" (Dowd, 1982, p. 443).

Eric Rasmussen takes another route to explain the ambiguities caught up in the idea of the moral hazard, warning that moral hazards sometimes bleed over into other market failures, such as adverse selection problems (Rasmussen, 2001).[5] As if underscoring the point, conservative commentator William Safire writes at times as though moral hazards are not adverse selection problems but instead a variant of the free-rider problem (Safire, 2003). By contrast, Brook Harrington invokes a different market failure to explain the moral hazard, characterizing moral hazards in terms of perverse incentives (Harrington, 2001). Bengt Holmström appears to believe the problem with the moral hazard is not a matter of free-riding or perverse incentives but a problem of asymmetrical information (Holmström, 1979). It is my view that these various manifestations are not simply a function of one singular problem but rather of multiple moral concerns intercalated into one idea (Hale, 2009).

There is, thus, a range of characterizations of the moral hazard, particularly among those who use the term most frequently, though few economists have stopped to outline the true meaning of the term or even its applicability to moral theory. For this paper, I would like first to pick up three representative positions that reflect tensions within the moral community between consequentialist moral theory, deontological moral theory, and virtue ethics. Later I will extrapolate from these positions to get a clearer sense of the arguments that one might encounter in the geoengineering discussion. I pick the first three only because they each emphasize normative theoretical dimensions of the moral hazard that otherwise rest just below the surface.

E. J. Faulkner explains that the moral hazard is "the intangible loss-producing propensities of the individual assured" (Faulkner, 1960). Steven Shavell, somewhat more concretely, defines the moral hazard as the "tendency of insurance protection to alter an individual's motive to prevent loss" (Shavell, 1979, p. 541). John M. Marshall proposes that the "moral hazard is commonly defined as excessive expenditure due to eligibility for insurance benefits" (Marshall, 1976). From these explanations, we can discern at least three characterizations of the moral hazard: what I shall refer to as the efficiency view, the reasons view, and the vice view.[6]

Faulkner's "efficiency view" emphasizes losses and, thus, the consequences of the action. Faulkner's view might be interpreted to suggest that the loss-producing propensities of the insured individual lead the agent to act *inefficiently*. This position appears to offer little more than lip service to the possibility that a moral hazard has a moral component. If one can be said to be violating some moral norm or principle, it is the principle of efficiency. Though it is true that many view inefficiency itself as a moral offense or, more generally, that many consequentialist and welfarist doctrines can be boiled down to claims about the moral undesirability of inefficiency, it is hard to see what is especially moral about *moral* hazards. They could just as easily be characterized as simple inefficiencies, apart from any claim about their morality.

Shavell's "responsibility view" proposes that what is wrong with the moral hazard is that it creates the conditions for insured individuals to ignore or disregard countervailing reasons. On this view, what is wrong with the moral hazard is that it alters the *motivation* of the agent and, thus, the reasons of the agent for acting.

Marshall's "vice view" offers yet a different objection. In this case, it appears that what is wrong with the moral hazard is that it encourages parties to engage in overindulgent (e.g., undesirable, negative, naughty) behavior, suggesting, thus, that temperance and prudence have fallen by the wayside. My suspicion, in fact, is that this is something like the commonsense view of the moral hazard. If this is the case, the value element implied by the moral hazard is a matter separate from the assessment of its alleged wrong. Society would have to agree on what qualifies as undesirably overindulgent and then clarify that the insurance situation will bring this overindulgence about. It may well be that there are reasons to refrain from excessive driving, for instance, but if there is a viable geoengineering solution to the climate problem, then those reasons cannot appeal to the negative impacts on the climate.

Moreover, the vice view is fraught with problems related to the benefits of insurance. One could just as easily argue that there are equally as many moral *safeguards* with insurance.[7] Street lamps in dangerous neighborhoods produce incentives for rogues *not* to mug or kill. Health insurance for babies *encourages* parents to take them to the doctors. Excessive consumption of insurance does not *necessarily* pose a moral problem at all. For another thing, this creates difficulties for ballyhooers of the position who appeal to moral hazard logic as a reason to abandon public programs. If it is the case that public programs ought to be abandoned, not because exposing oneself to more risk is morally problematic but because of the undesirable bad that the moral hazard brings about, then those who argue that the provision of insurance brings about the undesirable bad must argue not that *insurance* is the problem but instead argue for the moral impermissibility of the action taken to excess.

AMBIGUITY: THREE VARIANT MORAL HAZARD ARGUMENTS AGAINST GEOENGINEERING

Few of the commentators who link moral hazard arguments to geoengineering go into great depth on the line of reasoning that supports their concerns. The argument is typically presented as an aside, a quick-and-dirty way of suggesting that geoengineering has a moral dimension. From its mere mention, commentators appear to presume that the term "moral hazard" describes an ethically undesirable phenomenon. One problem, first and foremost, is that the characterization of moral hazards from geoengineering is ambiguous. Not only can it be viewed through the three moral lenses I discuss above, but it also seems to pick out three separate phenomena related to behaviors, beliefs, and counterfactual states of the universe. Yet, despite these complications, several have noted that it is one of the most compelling arguments against geoengineering.[8]

Above, I addressed how the general idea of the moral hazard might be viewed through the lens of efficiency, responsibility, or vice, any of which reflects a strong tradition within the ethics literature. In this section, I cover concerns of ambiguity related to three specific manifestations of the moral hazard argument against geoengineering. I term these the business as usual (BAU), counterfactual trajectory (CFT), and perverse behaviors (PB) variations. Importantly, any of the variant moral hazard arguments that I detail below could be viewed through the lens of the three variant approaches to moral theory—efficiency, responsibility, or vice—thus yielding *at least* nine variant interpretations of the geoengineering-related moral hazard argument. I actually identify sixteen. Consequently, in the section that follows this one, I return to these moral lenses and extrapolate from the below-discussed variant moral hazard arguments to build a much wider list of hazards and objections. I use this extrapolation to explore issues of vagueness.

To begin, one possible source of ambiguity rests in the isolation of the causal origin of the purported moral hazard. Some seem to suggest that the actual deployment of geoengineering will spur a moral hazard. If a geoengineering technology is successfully engaged, then this generates some morally perfidious outcome. Others seem to suggest that the mere *possibility* of successful deployment will spur the moral hazard. The mere knowledge of a viable Plan B will encourage shifts in behavior: just as fireproofing a house might lead one to expose oneself to more risk, so might knowledge that a house is insured against fire lead one to expose oneself to more risk.[9]

Below I identify three variations on the moral hazard argument and try to demonstrate why these variant moral hazard arguments are more ambiguous than clarificatory. Though each may at first seem similar, the moral hazard arguments engendered within are actually slightly different. Each calls attention to unique features of behavioral changes associated with geoengineering, and each suffers from both epistemic and justificatory complications. Consider the following three possible variations of the moral hazard argument. I touched upon each briefly in the introduction, but I would like to spend more time dissecting them here.

> **A. Business as Usual (BAU) Variation:** Geoengineering will make it possible to continue with BAU without any change in our (collective) behavior.

This, for instance, was David Keith's view mentioned above. It also is the view articulated by Arun Gupta, who notes that geoengineering may encourage the "continued use of oil, coal, and natural gas" (Gupta, 2010).[10] The hazarded wrong, in this case, is that behaviors associated with BAU will continue unabated. In other words, the underlying presumption of the BAU variation is that BAU is wrong or morally perfidious in some way. But there are at least two problems with this position. First, if the wrong of geoengineering consists in its permission of BAU, then there must be some attendant wrong with BAU. Most commonly, the wrong of BAU is presumed to be that BAU creates bad outcomes like climate change, though it is also conceivable that the wrong of BAU rests in a vice or an abdicated responsibility. If the wrong of BAU is fundamentally that the outcomes will be bad—that, say, BAU is causing climate change—and geoengineering changes those outcomes so that they are no longer bad, then the wrong of BAU would seem to disappear. So the BAU variation rapidly falls to tatters.

Second, despite its apparent wrongness, there are likely many positives to BAU. Burning fossil fuels, after all, brings many benefits, powering most of modern civilization. If we do geoengineer, this is the reason that we will have geoengineered in the first place: to *continue* with BAU without incurring the downside impacts of emitting. All SRM technologies are aimed primarily at avoiding only the downside impacts of climatic change, leaving in place many other problems related to land use, atmospheric carbon concentration, and ocean acidification. To suggest somehow that there is a problem with continuing with BAU requires an extra argument about why doing so is problematic. I will say more about these objections when I cover the objections and hazards below.

Consider instead a slightly different variation:

> **B. Counterfactual Trajectory (CFT) Variation:** Geoengineering will inevitably become part and parcel of BAU—Plan B will become Plan A—as we use geoengineering to offset our concerns about climate pollution.

This appears to be the moral hazard as explained by Samuel Thernstrom of the American Enterprise Institute, who describes the moral hazard as "the idea that greater consideration of geoengineering's feasibility might lead people to conclude that it is a viable alternative to emissions reductions."[11] The hazarded wrong in this case is not that BAU will continue unabated, but that *geoengineering* will be adopted into BAU. This concern, if it is a concern, rests on the presumption that a better state of affairs includes a world in which geoengineering is avoided but where emissions reductions are achieved. In other words, the presumption of the CFT argument is that there is some associated wrong with *geoengineering*, not with BAU per se, or that there is some associated right with a state of the universe in which we reduce emissions (Preston, 2011).

Put differently, the claim works by suggesting that an otherwise plausible counterfactual trajectory is the desirable trajectory. Along some counterfactual trajectory of the universe—a trajectory in which there is no Plan B—we would certainly have changed our behavior, maybe by conserving more or emitting less, but because there is now a Plan B, we no longer have a reason to change our behavior and, thus, can be said to increase our exposure to risk. The good is in the CFT; the bad is in geoengineering+BAU.

The first problem with the CFT variation is clearly epistemic. We do not and can never know what the counterfactual trajectory of our behaviors would have been, so we cannot say with accuracy whether humans would actually have changed their behavior. The second problem is justificatory. It is precisely because we acknowledge the difficulties or costs associated with CFT (conservation, for instance), that we are driven to seek geoengineering in the first place. The geoengineering is justified precisely because changing our behaviors is too costly, whether in expense, effort, or time. As a consequence, the CFT variation requires a non-question-begging argument that illustrates why geoengineering itself is an impermissible course of action. Such an argument cannot appeal to the counterfactual state of affairs because doing so would simply stipulate that that state of affairs is a better state of affairs.

Now consider a final variation:

C. Perverse Behaviors (PB) Variation: Geoengineering will entice us to behave in ways that are different than those that we have behaved in the past, and we may thus change our exposure to risk.[12]

Martin Bunzl seems to interpret the argument in this way. He writes: "Moral hazard only arises for geoengineering if you think that research or, if it came to it, implementation would undermine other actions and lead to *more not less* [emphasis added] greenhouse gas output" (Bunzl, 2009). The PB variation suggests that the associated moral hazard is not simply a matter of permitting BAU, but rather of exacerbating greenhouse gas (GHG) emissions either from what they currently are or from a hypothesized future state of the universe. The hazarded wrong in the PB variation implies that we well may be drawn to engage in behaviors that we might not otherwise. Perhaps we will increase our fuel consumption or our resource depletion; and perhaps these are undesirable or impermissible behaviors for other reasons. In other words, the presumption of PB is that there is something wrong with these new and encouraged behaviors.

Again, we face several problems if this is our interpretation of the moral hazard. The first is epistemic: there will certainly be pressures to increase our exposure to risk, but it is not clear that increasing our exposure risk by engaging in these behaviors is necessarily wrong. It may well be that geoengineering frees up money to build schools and hospitals, for instance. And if this is the case, it is hard to see where the problem is. Like eye wash stations in chemistry labs that serve as a constant reminder of the dangers of some chemicals, having an emergency solution on hand may offer the nudge into more responsible behavior that some environmentalists have been hoping for.

The second problem is again justificatory. If our concern is the instigation of perverse behaviors, then our reasoning needs to be checked. There must be some other feature of these behaviors that is problematic. If SRM enables us to build more hospitals and farms by offsetting the impacts of our carbon emissions, where is the wrong? If there is some other feature of these behaviors that is problematic, then we should be arguing against these problematic behaviors—that there are also attendant concerns about ocean acidification, land use, and air pollution—and not against geoengineering in general. The perverse behaviors variation simply asserts that SRM will permit these perverse behaviors to persist without offering a clear reason as to what makes these behaviors perverse.

To further complicate matters, all three of these variations do no work to specify what exactly the hazarded behavior is. In other words, each of these could specify myriad hazards, any of which might call into question the wisdom of a geoengineering scheme. The unique hazard, in other words, is vague.

VAGUENESS: HAZARDS AND OBJECTIONS

Now I would like to assess several related objections that could be alleged to inform concerns over moral hazards, whether of the BAU, CFT, or PB sort, and pair them with the moral hazard interpretations advanced by Faulkner, Shavell, and Marshall. These were, again, the efficiency view, the responsibility view, and the vice view. Working along two axes of ambiguity—the moral theory axis and the variant interpretation axis—we can begin to extrapolate a grid of more specific hazards.

	Moral Theory		
	Efficiency	Responsibility	Vice
Variations			
BAU			
CFT		Hazards and Objections	
		(Discussed Below)	
PB			

Figure 7.1. Table of Hazards

Since we can see above that the alleged wrong is ambiguous, and that in all instances an argument for the wrong or bad is external to the moral hazard argument itself (Hale, 2009), we will need to explore the various hypothesized hazards and objections in order to determine where the wrong might be.

One thing should however be clear: many of these arguments are not necessarily associated with increased exposure to risk. Sometimes moral hazard objections are better addressed when kept distinct from the confusing language of market failure. These are, therefore, what I am calling concerns of vagueness as it is not clear whether these hazards ought rightly to be considered the hazarded wrong that is the express concern of the moral hazard—that is, a consequence of geoengineering resulting in increased exposure to risk—or whether these are simply problems that emerge as mere consequences of geoengineering. At the end of the description of each hazard, I will indicate whether the hazard applies to the BAU, the CFT, or the PB variation.

One quick note before I begin. What follows is a complex and finely sliced assessment of a range of moral hazard arguments. Though the list is by no means exhaustive, I intend it as a compendium of individual hazards, each of which deserves treatment in its own right. Hopefully this list will motivate further research.

Efficiency Considerations

1. **Governance Hazard:** Geoengineering may lead to diminished public support regarding harder-to-implement policy interventions (Royal Society, 2009).[13] Just as a medical intervention that might cure alcoholism could encourage people to continue drinking without concern

for developing an addiction, so too might geoengineering permit individuals to continue consuming fossil fuels without concern for the climate effects. Plainly, there may be many associated objections with consuming alcohol, in one case, or fossil fuels, in the other case, but to maintain persuasive force, such arguments must offer further evidence that there is a wrong associated with BAU. Applies primarily to BAU.

2. **Snowball Hazard:** By encouraging BAU (or also perverse behaviors), geoengineering actually increases the likelihood of a bad outcome (Bunzl, 2009). Just as protective helmets may inspire athletes to believe themselves to be withstanding minor hits without permanent damage though they may be doing all-things-considered greater damage, so might geoengineering encourage behavior that results in a worse outcome than it otherwise might. What must be demonstrated for this argument is that there is such a phenomenon: that the damage associated with BAU+geoengineering outstrips BAU alone. Applies to BAU (or PB: see Equipment Hazard below).

3. **Technical Dependence Hazard:** We may grow so dependent upon geoengineering that we can never do away with it because we will always need to keep on top of things. Just as hydroelectric dams obviate the need for alternative water distribution measures, just as a decision to make railroad tracks a certain width commits engineers to that width throughout the transportation infrastructure, geoengineering may become fixed as the technology upon which all parties depend. If we are ever to lose funding or the capacity to geoengineer, climate change could be abrupt and far worse than it might be if we were to allow it to unfold over decades. To maintain persuasive force, such arguments must explain why such technical dependence would be problematic. Applies to CFT.

4. **Equipment Hazard:** By encouraging perverse behaviors, geoengineering actually increases the likelihood of a bad outcome (Bunzl, 2009). Just as protective athletic gear may inspire athletes to hit harder and thereby take greater risks, so might geoengineering encourage behavior that results in a worse outcome. As with the Snowball Hazard, so must it be demonstrated empirically that the negative outcomes associated with PB+geoengineering outstrip BAU alone. Applies to PB (or BAU: see Snowball Hazard above).

5. **Militarization Hazard:** Geoengineering will enable or encourage individual nations or rogue, independent entrepreneurs to conduct large scale experiments without global consultation or permission, thereby exposing the world to more risk. Just as biotechnologies invite consideration of military uses, geoengineering invites consideration of hostile use or abuse of others. Several have discussed this concern, though often not in conjunction with the moral hazard (see, for instance,

Robock, 2008). Says James Fleming: "It is virtually impossible to imagine that the world's powers would resist the temptation to explore the military uses of any potentially climate-altering technology" (Fleming, 2007). To maintain force, such arguments must offer an account of why the militarization would be wrong. Without an extremely stringent moral theory, few moral theorists would claim that all military technologies are morally problematic. Applies to PB.

6. **Regulatory Capture Hazard:** Geoengineering may enable or encourage institutional or regulatory capture (Stigler, 1971), thus decreasing the political viability of better options. Like corn ethanol, which initially was touted as a biofuels fix but eventually spawned perverse industries, concerns here focus on the potential for newly cemented political structures that may capture the political machinery and restrict further movement. As with many of the efficiency considerations, such a phenomenon must be empirically demonstrated both with regard to the presence of the phenomenon as well as the amplification of negative outcomes. If negative outcomes cannot be demonstrated, then a separate argument is required that offers a reason why regulatory capture is morally problematic. Applies to PB.

Responsibility Considerations

7. **The Band-Aid Objection:** Geoengineering treats only the symptoms and not the root cause of the problem.[14] If the symptoms disappear, the problem is allowed to fester. Like an industry that dumps pollutants into rivers, anticipating all the while that cleanup will be less expensive than installing cleaner technologies, geoengineering introduces the possibility that we are not actually addressing the core moral wrong of climate change (Hale, 2011). This is essentially the position advanced by Meinrat Andreae of the Max Planck Institute for Chemistry. Says Andreae: "You're papering over the problem so people can keep inflicting damage on the climate system without having to give up fossil fuels."[15] Blogger Jisung, a Ph.D. candidate in Harvard's economics department, also frames it this way. He writes: "Others believe that doing so might create a moral hazard problem, dis-incentivizing the necessary emissions reductions that must occur gradually, beginning now."[16] What must be explained here is why we have a responsibility to reduce GHG emissions if there are no negative consequences. Applies to BAU.

8. **Responsibility Abdication Objection:** Geoengineering will enable us to avoid or abdicate ourselves of responsibility for our wrongdoing now. Like a criminal who steals property but then pays to replace it, geoengineering permits us to get away with our crimes without paying

the true price. The hazarded wrong in this case is that we will not hold ourselves responsible for our wrong actions; geoengineering essentially provides a "get-out-of-jail free card" (citing an objection from Greenpeace: Royal Society, 2009). This hazard differs from the Band-Aid Objection in that it turns on responsibility and not on the act of wrongdoing. Applies to BAU.

9. **Political Noise Objection:** Geoengineering is a distraction for those who hope to engage in other activities.[17] Like a treatment for HIV that distracts from the many millions of people suffering from HIV, geoengineering promises to divert our attention away from our responsibilities to consider our emissions and our land use. Daniel Bodansky, for instance, understands the moral hazard problem as one that might "detract from emissions mitigation" (Bodansky, 2011). Again, this objection turns on political responsibilities. Applies to CFT.

10. **Cheating Objection:** Geoengineering inspires in actors a way of "cheating" on their responsibilities without going through the difficult moral retooling process engendered in other approaches to climate change. Like a student who cheats through an exam, geoengineering enables the world to "get the right answers" without doing any of the hard work required to get there. David Victor and colleagues argue that it will encourage "governments to deploy geoengineering rather than invest in cutting emissions. Indeed, geoengineering ventures will be viewed with particular suspicion if the nations funding geoengineering research are not also investing in dramatically reducing their emissions of carbon dioxide and other greenhouse gases" (Victor et al., 2009). Applies to CFT.

11. **Free Riding Objection:** Geoengineering will encourage people to avoid paying the costs of their actions and instead to free-ride, which is a problem. They can continue acting as they have always acted and also reap the benefits of a stable, controlled climate. Like offering international aid to the poor and destitute, geoengineering may encourage some people to rely on others, essentially eliding their own personal responsibilities. David Keith has offered an interpretation something like this: "[Changing one's behavior in the face of geoengineering] is not really a moral hazard; it's more like free-riding on our Grandkids."[18] To be persuasive, however, such arguments must explain why a given group must pull its fair share. It thus must also include a theory of fairness. Applies to PB.

12. **Political Strategy Hazard:** Geoengineering will enable individual nations or rogue entrepreneurs to *take advantage of* one another. Just as thermonuclear and intercontinental ballistic capacity enable individual nations to treat one another strategically, so too does geoengineering run the risk that individual nations or citizens will be treated

strategically. Similar to the Militarization or the Regulatory Capture Hazards, the hazarded wrong is that it will entice people to behave badly toward one another or to begin manipulating one another. Applies to PB.

13. **Perverse Profits Hazard:** Geoengineering enables private companies to profit off of a known wrong, so there are perverse incentives for companies to push for a geoengineering policy that is otherwise morally suspect. Like dentists who distribute candy at Halloween, or prison companies that lobby for tougher sentencing laws, geoengineering creates perverse revenue streams. This argument requires some explanation as to what is wrong with profiting off of cleanup. This may spill over into the Regulatory Capture Hazard, though the Perverse Profits Hazard suggests that it is wrong to benefit off the misfortune or mistakes of others. Applies to PB.

Vice Considerations

14. **Extravagance Objection:** Geoengineering "is used as an argument against painful curbs on our extravagant lifestyle."[19] Like a glutton who eats candy without concern for his weight or for the impressions of others, only to have laser lipolysis or radiation therapy to remove the excess weight, geoengineering permits us to live extravagantly without considering the virtue or vice of our action.[20] The hazarded wrong is that we will either continue being vicious or perpetuate the vice of extravagance. As before, extravagance or gluttony must be demonstrated to be a vice, and it must be shown so without appeal to consequences. Applies to BAU.
15. **Hubris Objection:** Geoengineering is an act of hubris. Just as some in the deaf community argue that it is an act of hubris to restore hearing to deaf children, inasmuch as it deprives a deaf child of developing in ways that are beneficial to her and instead seeks to "fix" them (Paludneviciene & Leigh, 2001), so too does geoengineering engender this sort of moral hazard. In this case it is the geoengineers, not the subjects of the geoengineering, who will be driven to viciousness. Manipulating the earth's climate essentially places human geoengineers in the role of gods, and as such, there is a hazard that these engineers will come to believe themselves to maintain control of the climate. The moral hazard is engendered in instigating the vice of hubris, of encouraging us to adopt further approaches in a way unbecoming of a good person. Applies to CFT.
16. **Attitude Objection:** Geoengineering encourages us to shift our attitudes toward the world such that we see it as something we can dominate and control. Just as with the Hubris Objection, we may begin to

adopt objectionable attitudes such that we view ourselves as more powerful or more perfect than we are. The difference between the Hubris and the Attitude Objection rests in whether the wrong is associated with how we will be versus how we might have been (the Hubris Objection) or how we will be despite how we might have been (the Attitude Objection). The hazarded wrong is that we will wander the earth with vicious attitudes, not that our other attitudes will necessarily have been virtuous. Contrast this also with the other vice objection, the Extravagance Objection, which is a BAU objection and associates the wrong with how we are and may continue to be. This is the concern expressed in the U.S. Government Accountability Office Report: that geoengineering will "undermine political will" (U.S. GAO, 2011, p. 54)—in other words, that the public will adopt the wrong attitudes—though later in the report, the GAO suggests that the moral hazard runs the risk that resources could be diverted from adaptation (p. 67). Applies to PB.

The above list is obviously not exhaustive, though it is drawn from the basic conceptions of the moral hazard articulated in the previous two sections. Given its contours, it may well come close to exhausting the possible objections.

It should be clear that much of our determination of the moral wrong depends on what is meant by "moral" in moral hazard. If by moral we take the consequentialist line and reason that the best action is that action that maximizes the good for all, then insurance, when properly executed, is clearly the best possible option, even in spite of proposed moral hazards. If by moral we mean something akin to having the wrong character traits, or falling prey to vice, then we must make the argument for the moral failing. So too if one takes the deontological line and reasons that the best action is the action that functions according to a universalizable maxim, or to some principle of the right, then this maxim will remain steady regardless of consequences. The important point is that there are many possible hazards associated with shifts in behavior and captured under the "moral hazard" umbrella. Which of these hazards emerges as the core concern will depend in one part on the moral theory we apply and in another part on which of the several variant moral hazard arguments we invoke.

CONCLUSION

As we have seen above, there is no univocal moral hazard argument against geoengineering. Rather, there is a cluster of arguments, each related roughly to some ambiguous behavioral phenomenon that has been called the "moral hazard." As a consequence of this clustering, moral hazard arguments against geoengineering are beset with concerns of ambiguity and vagueness, not to mention accuracy. Confusion surrounding the moral hazard argument against geoengineering is a consequence of ambiguity about the very idea of the moral hazard. It emphasizes the behaviors (BAU), the geoengineering itself (CFT), or the increased exposure to risk (PB), and in doing so implicates this dimension of behavioral change as the dimension of ethical concern. Unfortunately, moral hazard arguments against geoengineering are not simply ambiguous. They are also vague. The vagueness of the moral hazard arguments rests in follow-on considerations regarding the alleged bads. As a consequence, all moral hazard objections require at least some further justificatory labor inasmuch as they are not self-explanatory bads.

Finally, it should not go without stating that all moral hazard objections raise empirical questions about accuracy. It is a fact of the world how people will respond to geoengineering, and so even if considerations of ambiguity and vagueness are addressed such that we finally gain a grip on the true moral hazard in question, we then must answer the question of risk: of whether the predicted outcomes are likely to come to pass. As mentioned earlier, the Royal Society Report recommends further research to analyze the extent of the moral hazard effect (Royal Society, 2009, p. 45). But no amount of empirical research will overcome the conceptual confusions associated with the various changes in behavior that geoengineering may provoke; and none will address concerns about the moral valence of these changes in behavior.

The problem with moral hazard arguments can thus be demonstrated. Inasmuch as moral hazard arguments are ambiguous and vague, they are easily sidelined and dismissed for any range of reasons: because they are deemed conceptually intractable, because they are thought to fall easily to simple retorts and replies, or because there is always more information that must be gathered to make the case. Consider Martin Bunzl's articulation of the moral hazard argument again: "Moral hazard only arises for geoengineering if you think that research or, if it came to it, implementation would undermine other actions and lead to *more not less* [emphasis added] greenhouse gas output" (Bunzl, 2009). I understand Bunzl's position to be either the Snowball or the Equipment Hazard. Bunzl then offers essentially the following dismissal of the Equipment Hazard: "Antilock braking systems and airbags may cause some to drive more recklessly, but few would let that argument outweigh the overwhelming benefits of such safety features" (Bunzl, 2008).

Bunzl's reply illustrates how convoluted the discussion on moral hazards can become and why responses to the variety of moral hazard concerns are rarely satisfactory. Antilock braking systems and airbags may result in more reckless driving, this is true; but it is not clear that they result in *more injuries* from increased reckless driving, which is why we continue to use technologies like antilock braking systems and airbags. In fact, reckless driving, or at least faster driving, may be the sort of thing that we want to encourage, precisely so that we can each get to our destinations faster. We could easily drive five miles per hour and be exceptionally safe, but such safety measures would be overkill. What must instead be specified is *where the wrong is*. It clearly does not lie in taking our cars out on snowy days or driving at high speeds on designated roadways. That is the reason, after all, that we continue to utilize antilock braking systems and airbags.

Clarifying which of these problems the moral hazard argument engenders will go some distance in dispelling ambiguity and vagueness. What each argument needs is treatment that attends not only to the phenomenon that individual or collective actors will change their behavior in the wake of a policy intervention, but some clarification of what is wrong with changing behavior in that particular way. My suggestion is to be more specific in outlining the concerns associated with geoengineering. The above analysis illustrates that there are many complicated moral questions associated with geoengineering, any of which must be explored in greater conceptual and empirical detail. Further analysis will hopefully guard against confusions.

NOTES

1. www.ted.com/talks/lang/eng/david_keith_s_surprising_ideas_on_climate_change.html
2. Used thusly, a *prima facie* reason is an "intrinsically moral," albeit defeasible and non-absolute, reason (Dancy, 2010; Ross, 1930; Searle, 1978).
3. Portions of this section have appeared elsewhere, though they have been substantially revised to address the concern of moral hazards in geoengineering (Hale, 2009).
4. Another market failure typically associated with insurance is the adverse selection failure, but this is unrelated to concerns over geoengineering. (An adverse selection failure occurs when those most likely to need insurance are those who take out insurance, as would be the case with high-risk cancer patients who might take out health insurance with special knowledge of their condition. This has the effect of distributing risk across a narrower body of individuals instead of across the entire population, thereby driving the cost of insurance higher.) Because geoengineering is a global insurance plan, adverse selection is not a problem.
5. "From the principal's point of view, agents are identical at the beginning of the game but develop private types midway through, depending on what they have seen. His chief concern is to give them incentives to disclose their types later, which gives games with hidden knowledge a flavor close to that of adverse selection" (Rasmussen, 2001, p. 241).

6. To be fair, none of these authors set to the task of specifying the normative force of the moral hazard. All three of these articles define the moral hazard in one sentence and then proceed to assess its economic implications in given situations. Still, I think these positions provide a nice starting point for this discussion.

7. In recent work, Deborah Stone (1999–2000) has characterized such moral safeguards as "moral opportunities." I am inclined to agree with her characterization, though her approach is to argue that "the act of participating in insurance can be and often is a highly moral choice, because (following another long line of thought), insurance is a form of mutual aid and collective responsibility."

8. Apart from the Royal Society Report, see also: http://scienceline.org/2011/04/turning-down-the-sun/

9. For ease I use the term "geoengineering" throughout to refer both to the actual deployment of such a technology as well as the possible deployment of geoengineering. Thus, by "geoengineering" I mean "geoengineering or the possible successful deployment of geoengineering."

10. Gupta writes: "Is talk of geoengineering creating a 'moral hazard,' encouraging the continued use of oil, coal, and natural gas because we can presumably counter the effects?" http://www.zcommunications.org/geoengineering-the-planet-by-arun-gupta

11. Thernstrom clearly does not agree that this is a serious concern. He writes: "The one argument that could still derail research proposals is a misplaced fear of the moral hazard—the idea that greater consideration of geoengineering's feasibility might lead people to conclude that it is a viable alternative to emissions reductions." http://www.american.com/archive/2010/march/what-role-for-geoengineering

12. This view was advanced by Josh Horton on the Geoengineering Politics blog: "people will embrace geoengineering as an excuse to avoid emissions reductions, and current levels of fossil fuel consumption will persist if not increase." http://geoengineeringpolitics.blogspot.com/2010/09/what-moral-hazard.html

13. The International Risk Governance Council raises related concerns: "The moral hazard is that a decision to support geoengineering technologies could lessen efforts to reduce the global concentration of CO2 and other greenhouse gases" (www.irgc.org/geoengineering). Also expressed by John Shepherd, September 3, 2009. New Scientist: www.newscientist.com/article/mg20327245.600-geoengineering-is-no-longer-unmentionable.html. Also in the Royal Society publication: September 2009: http://royalsociety.org/Geoengineering-the-climate/

14. Advanced in the John Martin Geoengineering Working Group paper (p. 2): www.practicalethics.ox.ac.uk/_data/assets/pdf_file/0013/21325/Ethics_of_Geoengineering_Working_Draft.pdf

15. www.essc.psu.edu/essc_web/seminars/fall2006/KerrGeoengOct06.pdf

16. www.senseandsustainability.net/2011/08/29/does-investment-in-geo-engineering-create-a-moral-hazard-problem/

17. "Geoengineering could also be perceived as a moral hazard, as there is the possibility that it could decrease the political and social impetus to reduce carbon emissions." www.eastasiaforum.org/2011/07/29/geoengineering-and-tackling-climate-change/

18. Oral Comments, Royal Society Conference, "Geoengineering the Climate - Science, Governance and Uncertainty," September 2009. Author's note: This quote has generated some consternation amongst the blogging community. It appears to have been taken out of context by some groups that are vehemently opposed to geoengineering. In my characterization of the quote above, I believe I remain true to the spirit of the quote. Keith seems to think that the core concern of the moral hazard objection is not simply that people will change their behavior but that they will essentially be free-riding on others.

19. John Nissen; October 15, 2008. http://groups.google.com/group/geoengineering/browse_thread/thread/fc1acf7aa46903f/4e5f90b0aa645d9b?lnk=gst&q=hazard#4e5f90b0aa645d9b

20. Joe Romm sometimes uses examples of this sort: www.nytimes.com/2011/10/04/science/earth/04climate.html?_r=1 and http://thinkprogress.org/romm/2011/10/06/336676/geoengineering-panel-climate-remediation/

Chapter Eight

Climate Remediation to Address Social Development Challenges

Going Beyond Cost-Benefit and Risk Approaches to Assessing Solar Radiation Management

Holly Jean Buck

Many people are averse to the idea of solar radiation management because it is seen as a technological solution to a social problem. Indeed, climate change is a social problem, as cultural, economic, and political dynamics are currently inhibiting the transformation of our fossil-fuel intensive energy and land use systems. One commonly cited concern with using solar radiation management (SRM) to combat global warming is that SRM could slow down or even preclude opportunities to transform present fossil fuel based energy and land use systems into sustainable and socially equitable systems.

However, it is possible that climate remediation strategies could also be social development strategies. Some technologies, such as increased crop reflectivity, could potentially be coupled with policies and be deployed in ways that would have direct benefits to societies, communities, and individuals. For other techniques, however, the possibilities of scaling them up and deploying them in ways that would maximize social co-benefits or help overcome socioeconomic inequalities might be inherently restricted.

This paper argues that assessing potentials and opportunities for socioecological change should be part of assessing SRM. An ethical framework for assessing SRM technologies would take into account whether or not the strategy could conceivably be scaled up in ways that alleviate structural aspects of human development problems such as inequality, energy poverty, food security, and land access. Factoring these potentials into assessments does not mean that the potentials will be actualized, of course—even a climate remediation strategy with such potential social co-benefits as afforestation is often done in a way that exacerbates land tenure issues and displacement—but this paper proceeds on the premise that it is important to recognize whether the potential for transformative deployment exists. If it is ethical to give future generations the power of choice, the amount of social choice the technologies provide should be a part of our assessments.

The idea of assessing climate engineering, though most everyone agrees upon the need to do it, is not entirely thought out yet. Still, there are numerous assessment efforts underway at present. The first section of this paper briefly discusses current approaches to assessment and identifies three conceptual challenges that make our current assessments inadequate. The next part of the paper suggests what a better assessment process might evaluate, resting upon two key ideas. Firstly, we should evaluate the amount of path dependency and lock-in, social and technological, that the technologies afford. Secondly, we must focus more attention upon identifying potential co-benefits that could address existing issues around ecosystem degradation and access to food, water, energy, and land.

ASSESSING ASSESSMENT: WHY CHOOSING HOW TO CHANGE THE CLIMATE IS NOT LIKE SHOPPING

Most parties agree that climate remediation technologies need to be assessed effectively, objectively, and carefully. In a worst-case scenario, smart assessment could be what keeps us from catastrophe. But "effective" and "objective" are slippery words. What else does an assessment need to be? Does it originate from models, does it precede research or come afterwards, is it timely or thorough, democratic or expert-led? How does it allow for feedbacks? What makes an assessment ethical, in content or in process? The entire idea of assessment is an optimistic exercise in many respects and perhaps a progressive one. It suggests that we, as a single rational actor, are able to evaluate technologies and make choices about their use.

Social convention (or academic and policy convention) offers the rational chooser many approaches to assessment. We have environmental impact assessments for events that affect the environment, technological assessments for new technologies, cost-benefit assessments for matters of economic weight, risk assessments for risky matters, vulnerability assessments for vulnerable populations, threat assessments for dangers, and so on. We tend to use the kind of assessments we have developed for other efforts, the familiar tools of our disciplines. All of these types of assessments are applicable to a global, crosscutting issue like climate remediation, but how do they get combined into a holistic view of climate remediation—and what do they leave out?

Assessment, by a dictionary definition, means to "evaluate or estimate the nature, ability, or quality of." Do we find it important to assess the ethical implications of SRM—what it means for humans to change the balance of incoming sunlight? Or are we primarily assessing our own abilities to engineer the climate—the feasibility of different strategies? If we are assessing the quality of these strategies, are we talking about economic quality, environmental quality, social quality? Rarely, if ever, are social structures or opportunities linked with climate remediation. Hence, there are numerous questions of the content of the assessment (that is, whether the assessment lists techniques and rates their feasibility, or whether it delves into principles of research, governance implications, etc.).There are also related questions of process—how the assessment is conducted, who is doing the assessing, what their values are. This paper does not focus on the process of doing these assessments, though that is a worthy topic (see Corner & Pidgeon, 2010; Stirling, 2008). Rather, this paper is interested in the metrics we are using to assess climate remediation and whether these metrics give us the best results, given a goal of a just, climatically stable, and sustainable world. Let us briefly survey the criteria used by current assessment efforts and then look at the limitations of these criteria.

Several transdisciplinary assessments of climate engineering have been recently completed by science and government organizations, such as the UK's Royal Society (2009), the Netherlands Environmental Assessment Agency (2009), and the U.S. Government Accountability Office (GAO) (2011). The Royal Society report has become a central referent in almost all climate engineering discussions. Its diagrams that plot geoengineering strategies along axes of effectiveness and affordability, with color-coded circles designating safety and timeliness, are familiar to most who have ever watched a power-point on this topic. Similarly, the U.S. GAO rates technologies on basically the same four categories used in the Royal Society report: maturity, potential effectiveness, cost factors, and potential consequences. However, it also features a survey of public attitudes toward geoengineering. The Netherlands Environmental Assessment Agency report uses the four criteria of effectiveness, feasibility, environmental implications, and political implications—the latter of which goes beyond the Royal Society's criteria. Another set of criteria was created in a public dialogue conducted by the UK's Natural Environment Research Council, which recommended assessing technologies based on controllability, reversibility, costs and benefits, timeliness, and potential for fair regulation (Godfray, 2010). The most global and comprehensive transdisciplinary assessment process is the Intergovernmental Panel on Climate Change (IPCC) assessment report, the fifth of which is being developed now for a 2013 release (AR5). Working Group I (WGI) will assess the physical science basis of geoengineering, WGII will assess the impacts on human and natural systems, and WGIII will "take into account the possible impacts and side effects and their implications for mitigation cost in order to define the role of geoengineering within the portfolio of response options to anthropogenic climate change" (IPCC, 2010, p. 1). It will also look at options of appropriate governance mechanisms.

These reports, along with several other studies with disciplinary approaches (Goes, Keller, and Tuana, 2010; Lempert & Prosnitz, 2011; Russell et al., in press), have generally created common metrics of effectiveness, timeliness, cost, and safety. Let us leave aside for the moment the full story of *how* these metrics might be becoming a standard way of evaluating interventions—we could visit Descartes, risk society, neoclassical economics, and other thinkers and concepts from the last four centuries along that long road. For now, we will simply discuss the fundamental problems with the limited metrics and conceptualizations we are using to evaluate SRM. There are three interrelated problems with our assessments that I would like to highlight in this section: firstly, the problem of the common human actor; secondly, the rift between the natural and social worlds; and thirdly, the focus on side effects and impacts.

Our first conceptual stumbling block is the "prudential framing" identified by Gardiner, who explains that the Royal Society report "implicitly sets out the climate problem as a prudential one facing a single agent, humanity" (2011b, p. 7). This is problematic because humanity is not just "one big, happy, global and intergenerational family" (Gardiner, 2011b, p. 7) whose members share the same interests. Gardiner argues that a successful account of humanity's collective interests "needs to reconcile interests that often seem at least modestly at odds (e.g., those of the current rich and poor, those of current and future individuals, and so on)" (Gardiner, 2011b, p. 7). Assessment frameworks that look to maximize benefits and minimize costs or that assess risks and impacts make unspoken assumptions about what this common human actor wants and what the values underlying his or her rational choices are. Moreover, even when we perform a scientific assessment that offers this common human actor the power of informed rational choicemaking, real people might not take the rational choice because of the morass of other influences, which might include narrative, cultural, political, and financial. Those "hidden" influences are also forces acting upon the complex system. A more effective, honest assessment process would make them visible.

A second conceptual challenge with prevalent SRM assessment metrics is the fragmentary disciplinary lenses from which these metrics have arisen. We need every discipline to understand climate remediation. Yet despite nearly every report's recognition of the need for social science, ethical, and humanities research, the social is still an undeveloped and under funded shadow compared to the scientific and technical realms. You can see the authors of the Novim report grapple with this:

> The research agenda presented here focuses only on the engineering and technical aspects of these questions. . . . This study has focused only on the question of what technical research agenda could be pursued to maximally reduce technical uncertainty in climate engineering responses to climate emergencies. *Socio-political factors and challenges were explicitly sidelined in order to facilitate an unconstrained assessment of the technical aspects* [emphasis added]. However, socio-political issues clearly cannot be ignored in— and in fact should be central to—any decision making process about climate engineering research and interventions. (Blackstock et al., 2009, p. 30)

Unfortunately for report writers, it is impossible to completely sideline socio-political factors and challenges when considering geoengineering technologies. They are inseparable. The socio-political questions are not just for "decision makers" and "social scientists"—every natural scientist is also a decision maker, deciding what to research and how to go about it, what questions to ask.

Policymakers and scientists are aware of the challenge of integrating, or at least balancing, social and natural sciences within assessments. Even though the Royal Society report divides the criteria by which geoengineering proposals should be evaluated into two classes, "technical criteria" and "social criteria," the report spends two chapters upon a "preliminary exploration" of social considerations. It advises that geoengineering proposals should be assessed against non-technological criteria, which "include issues such as public attitudes, social acceptability, political feasibility and legality, which may change over time" (Royal Society, 2009, p. 21). Recognizing the need for more sophisticated assessments, two of the UK's national research councils have funded the Integrated Assessment of Geoengineering Proposals (IAGP). This project looks into geoengineering broadly, using climate models and public dialogue in which "the public will contribute to the criteria and metrics formed to evaluate geoengineering proposals" (Parkhill & Pidgeon, 2011, p. 4).

The disciplinary challenge is only a small part of the Cartesian binary: the split between mind and body, human and natural, social and environmental. This divide is intimately related to the third conceptual challenge we face in creating better assessments, which is the mechanistic emphasis on impacts and side effects. This emphasis comes from a way of thinking that imagines humanity as a force that acts upon the world, producing impacts and side effects. Is this a useful way to see the world and our place within it? This chapter argues that it is not, for a few reasons. One is that this worldview is premised upon the Cartesian binary: it assumes that humans are not a part of the rest of nature. Another is that it is too simplistic of a worldview to be useful for our current socioecological challenges.

Consider the premise that climate remediation strategies, or methods or technologies, are not *just* technologies. Neither are they activities that occur within the bounded terrain of "the environment" nor paths of action that happen and have "impacts" or "side effects." Rather, climate remediation strategies would create new coupled human-environment systems. They would create ways of ordering human-nature relations that are both ecological and social. Social structures are part of the package when you pick a climate engineering technology from a neatly ordered diagram. Hence, climate remediation actions should be evaluated beyond assessments categorized into technology, science, environment, or impacts. They are more than just the application of scientific knowledge for practical purposes, and they exist beyond the environment.

To see why this is so, let us look at some examples of other technologies that are involved in coupled human-environment systems. Think about the strategy of extracting oil from the Athabasca tar sands. The *in situ* technologies of applying solvents or steam to heavy bitumen are an application of science and engineering genius for a practical purpose, generating energy usable by the current fossil fuel infrastructure. (They also have another purpose, namely, to generate revenue, but that is beyond this discussion.) When put in a matrix of effectiveness, cost, safety, and timeliness, would tar sands oil do well compared with other means of producing energy? This depends upon what kind of "externalities" are factored into the cost (such as additional CO_2 emissions, water usage), how safe the climate impacts are considered to be, the ways in which one factors in the social harm to indigenous communities, etc. In addition, tar sands extraction certainly would not win any merit for elegance, if that were a criterion of the assessment.

Extraction technologies, or energy technologies in general, have traditionally not been subject to the kind of broad four-category assessment described above. It would certainly be an improvement if they were, as the economic and environmental absurdity of projects like the tar sands would be revealed through the act of comparing them to other technologies. Still, if energy technologies were subject to broad, comparative assessment using a four-criteria matrix of effectiveness, timeliness, cost, and safety, it would leave out important human, social, and political dimensions. The tar sands are not just about technology; they are about North American oil companies, about the geopolitical calculus of where North America gets its energy, and about narratives of jobs and security. They are about lock-in dependence upon dirty oil, about increased incidence of cancer, about First Nation politics, about what people in First Nation communities have termed "ethnocide." Turn to another country, and the complexity of the socioecological calculus is also visible. Nigeria faces armed groups attempting to gain a just share of the Niger Delta's oil wealth; Azerbaijan is experiencing an oil boom that has both lined the pockets of the Aliyev dynasty and funded the modernization of Baku. The application of technology can create class stratification, corruption, and explosive social dynamics. It would be very difficult to include quantitative measures of these effects in an assessment—but it would be an enormous mistake to ignore them on the basis that it seems too complicated.

Another salient example of a technology that was not well-assessed is ethanol. This energy strategy was excitedly championed by the farm lobby, resulting in massive land use change—in the United States, the Department of Agriculture tracks that around 40 percent of domestic corn use at present goes toward ethanol. This contributed to a spike in food prices, especially maize—with painful consequences in nations like Mexico where maize is a staple. This arguably should have been part of the assessment on whether to make generous ethanol subsidies into policy, as the side effect of people not being able to feed themselves is important.

To reiterate, the socioecological changes that climate remediation could cause should be assessed beyond how they are "side effects," because these social structures are more than a side effect of a technology. They are the fabric of this world. Climate remediation would not just "impact" this fabric; it would be woven into it. Industries would grow up around cloud whitening ships or albedo modification schemes, creating livelihoods and shifts in local job markets, potentially changing architectural designs or identities of places. Whiter skies and redder sunsets would work their way into cultural forms, becoming symbols of the change that has taken place. If built, a solar sunshade could become a mythic project much as space travel was in the mid-twentieth century: an image and concept with symbolic resonance. Climate remediation technologies could come to connote peace or threats and potential war, depending on how they were developed and deployed. All of these technologies could inspire people to think differently about humanity's place or role in earth systems, and that conceptual change would catalyze other projects and ways of organizing society. Accepting responsibility for managing solar radiation could transfer to accepting other responsibilities, some of which are already in different stages of management: agriculture, fishing, mining, energy production. None of these changes in the fabric can be quantified; the point is that they are not "side effects," but integral parts of a complex system.

As part of a reflexive human-environment system, technologies presumably respond to that which they are influencing. With the example of the tar sands, one could imagine that the production and combustion of that oil would release enough carbon that some legislation finally gets passed which makes the tar sands too expensive to exploit, and then the technology would have to be abandoned or modified to somehow make the extraction process cheaper again. With this boom-and-bust cycle come changes in demographics, political views, economies, and ecologies. Or in another context, perhaps the threat of terrorism from Niger Delta militants reaches a level that investment fails completely and oil companies withdraw. Or perhaps democracy works in Nigeria, and the government manages to build a sovereign wealth fund that manages the revenue to grant healthcare, education, and infrastructure for the people. One can imagine that with ethanol subsidies, international pressure from Mexico and other nations regarding the price of maize allows U.S. domestic agricultural policy to change—but in order for the corn lobby to accept the loss of subsidies, other concessions have to be made in complex legislative negotiations, which result in different land use and different ecologies down the road.

The reflexivity of these complex adaptive systems has been pointed out by Allenby, who explains that a climate change model does not change if it is run; it has given parameters it uses throughout. However, "when the results of the model run are published and presented to policymakers, who then change their policies in response, the assumptions of the model may become dated, or even at some point dysfunctional" (2010, p. 10). He continues, "[T]he model has been undone not by any internal failings, but by the reflexivity of social and cultural systems." There is a double lesson here. One is that the technologies we are assessing are part of reflexive, complex adaptive systems, and they will do more than just impact the system with side effects—but change it in ways that will reverberate through the whole socioecological system. The technologies do not affect one specific domain (the climate, or the temperature) but entire socioecological systems. The second lesson is that the *assessment process itself*, and thus the criteria used in it, may have socioecological repercussions. This is a crucial point, because it means that the assessments we write do not remain in the realm of the social: they can influence and produce specific ecologies.

A more sophisticated, cross-cutting assessment would take into account how those socioecological systems could change. It would take this into account not simply in a reactive way but in a way that allows for human decisions and inputs, for human agency. In the IPCC Special Report on Emissions Scenarios (SRES), families work toward this to some extent, as their story lines take into account worlds with differing economic and environmental values and approaches to globalization, but they do so in such a way that only *implies* human agency. This kind of assessment is perhaps something we have never done throughout our history of coming up with new ideas that shape our world—though it is perhaps something we should have been doing.

WHAT DOES A BETTER ASSESSMENT PROCESS LOOK LIKE?

Thus far, this paper has identified many opportunities for improvement in geoengineering assessments. A better assessment process would be interdisciplinary; it would be sensitive to the many intentions, cultural perspectives, and agendas beyond the prudential; and it would address coupled human-environment systems. This section of the chapter expands upon two key aspects of a better assessment process: evaluating climate remediation ideas based upon the amount of social choice they afford, and evaluating them based upon the social and ecological co-benefits they confer. Because assessments guide research priorities, these two metrics move toward research that would support socially transformative deployment, that is, climate remediation techniques that could also promote social justice.

Transformative deployment is based upon the idea that our ills are not simply environmental but also social. Hence, we need solutions that restore both ecosystems and societal structures. Transformative deployment also suggests that we need to be more forward-thinking: climate engineering is about actions that we are taking. Whether it is conceptualized as designing, engineering, intervening, managing, or problem-solving, these are all active processes. Yet geoengineering is somehow still commonly constructed as *reactive* even though it is about committing an action. A consequentialist, utilitarian focus on impacts—such as, we do x, then y place is impacted in z ways—continues a conceptual framework that is not proactive.

If we accept that climate engineering is not simply reacting to climate change but taking on our systemic food, energy, and climate challenges, we could make stronger advances toward addressing these challenges. Rather, we could think: we want these z characteristics in our climate, agricultural, and energy systems, so what x action do we need to do to attain them? Is it possible to imagine a more sustainable and socially just world *first* and *then* determine what engineering strategies could help get us there—rather than starting from the technologies and modeling what might happen with them? For if we have accepted the idea that we must now design features of the environment, we should put socially just intentions into our designs. Stabilizing the climate with geoengineering could be an ethical action to take; it could help people suffering from climate change. Yet even with a relatively stable climate, at present millions of people cannot access food and water due to distributional issues. Being proactive about climate engineering would mean consciously trying to do better than the current status quo. We are understandably afraid of the "hubris" involved with living by design, yet it is quite possible that a reluctant engineer would do a worse job than a forward-thinking one.

It is bold to begin with a target vision because we cannot decide what world future generations will want and because there is no single, agreed-upon human vision of a future world. Here we face the limitations of the "common human actor" again. Postmodernity and the specters of totalitarianism effectively obliterated the possibility of there being a single human actor. As Humphreys notes, the goal might appear to be to return the temperature to a pre-industrial "natural" state, but land use has changed significantly since then, and he predicts that scientists and politicians would disagree on a desired mean temperature (Humphreys, 2011, p. 111; see also Jamieson, 1996). There is no common human actor to agree that the pre-industrial temperature would be the ideal one to return to. Hence, one next best thing we can do for ourselves and for future generations is to pursue strategies that offer future generations the maximum amount of choice. An ethical SRM assessment would take the availability of choice into account. This type of assessment would be grounded in the idea of affording intergenerational equity—which, as Burns describes, is a principle of distributive justice. A principle incorporated in the United Nations Framework Convention on Climate Change, distributive justice demands that present generations not take benefits at the expense of future generations (Burns, 2011, p. 41, following Maggio, 1997). In Burns's analysis, "deployment of SRM geoengineering technologies in the future could constitute the quintessential act of generational selfishness, compelling untold future generations to 'stick with the program' or face catastrophic impacts," because of the potential moral hazard, collateral effects, and the termination effect (2011, p. 55). Are there strategies that could place future generations in a better position? To minimize intergenerational threats (or to investigate if it is even possible to do so), a first step would be to assess strategies on the basis of the level of choice of social-ecological systems they afford, thus hopefully enabling them to avoid societal and technological lock-in.

EVALUATING OPPORTUNITIES FOR SOCIAL CHOICE

When it comes to solar radiation management, many are worried about technological lock-in. SRM approaches such as stratospheric aerosols and space-based reflectors could have a devastating "termination effect," where the cessation of SRM would cause very rapid warming that ecosystems would have a more difficult time adapting to. This could potentially cause crop failures, rapid climatic changes, and permafrost thaw (Royal Society, 2009, p. 35; Matthews & Caldeira 2007; Russell et al. 2011, p. 15).This technological lock-in has certainly been noted in major climate engineering assessments. However, we should also pay attention to and seek to avoid lock-in not just from the nature of the technology, but from the social structures to which the technology becomes coupled.

An obvious example of devastating social lock-in is the fossil fuel system that got us into this predicament. As Lohmann points out, fossil fuels "weren't chosen because they were a rational, low-cost, efficient means of meeting pre-existing ends, but for other reasons" (2006, p. 112). Timing and historic circumstance, political choices such as subsidized structures for high fossil fuel use, the triumph of alternating current over direct current, and other social and political processes shaped the possibility of talking about fossil fuels as "cheap" or "efficient" today (2006, p.111). If fossil fuels had been evaluated for their social lock-in and the enormous monetary, ecological, and social burdens they would have caused, we might well have decided to use them anyway, given the benefits they have—but we might have gone about it differently, reducing lock-in by setting up different ownership and use arrangements. Or maybe our hypothetical rationally-choosing governance regime of the past would have chosen never to go down this road. Regardless, comprehensive assessment with attention to social lock-in likely would have drawn attention to the potential problems.

Any coupled natural-social system will exhibit some degree of social lock-in. Take, for example, the strategy of afforestation. Large palm plantations have recently grown up around the world, converting smallholder agriculture or forest into monoculture. This system will have some social inertia because the land has been converted, corporations have already invested, and people are already displaced. If these plantations are created for carbon credits under the clean development mechanism (CDM), it also contributes to locking the global south in the role of providing raw materials for the global north and does nothing to change the social structures for emitting nations in the north. Given this, how is it possible to evaluate how much social energy would be required to change this social arrangement for another one? An economist could probably come up with a dollar value, which would be an imperfect measure. However, for afforestation, the social lock-in is in the design; there would be ways to implement afforestation that did are more flexible—beyond palm monoculture, using other schemes that are more development-oriented and friendly to smallholders.

SRM strategies should also be evaluated on the amount of flexibility they have in being deployed in ways that avoid both technological and social lock-in. One key difficulty in this is that what one actor sees as a "social lock-in" might be viewed by someone else as a "constant revenue stream." Thus, lock-in might be difficult to avoid. There are entire marketing departments worldwide pursuing social lock-ins of one sort or another. Yet, returning to the idea of giving future generations the most choice possible, it is ethical to identify both social and technological lock-in of technologies.

Could the risk of social lock-in could be rated high, medium, or low, much as the Royal Society report rates the risk of the termination effect? This requires further exploration. For example, consider stratospheric aerosol injection schemes. They would have a strongly negative weighting when it comes to technological lock-in, but it is not so clear for social lock-in. Deployed technocratically, under the discretion of a select few, one could argue that the social lock-in is not so great, as only a few actors would be involved in implementation. Deployed democratically, it is also possible that the social lock-in is not great because the possibility for change exists. Social lock-in stems from having a magnitude of dispersed actors (e.g., coal and oil consumers), but also from a few vested interests with a concentration of power (e.g., fossil fuel companies). It is developed more through the design of the deployment and governance rather than inhering to the technology itself. Hence, an assessment might not want to attempt to give a numeric or qualitative value to the prospects for social lock-in; rather, it could name and identify the opportunities for social choice. The process of assessing would make clear how important the design of the deployment and governance would be. The assessment of social choice would be a complement, but not a replacement, for the scientific evaluation of technological lock-in (as in the termination effect for stratospheric aerosols).

IDENTIFYING CO-BENEFITS

An idea that follows from avoiding lock-in is to assess what potential co-benefits strategies offer. This can be another form of offering choice to the future: first, the choice to not be locked into a certain social and technological regime; and then, more expansively, the choice to arrange SRM deployment to maximize co-benefits.

This is an idea that is just beginning to be explored. In a scoping workshop done by the UK research councils, researchers identified examining co-benefits as a major opportunity for research with food security, energy security, economics, and ecosystem services all mentioned as potential co-benefits (2009, p. 16). The recent report by the Wilson Center, *Geoengineering for Decision Makers*, suggests that "both geoengineering and climate change should be addressed in the context of finding 'simultaneous solutions' to other challenges, such as energy security, vulnerability to terrorism, water scarcity and food security, ocean health, economic competitiveness and job creation" (2011, p. 3). It is increasingly clear that we need to identify and expand upon synergies between the ecological and social restoration needed. Our review of the co-benefits for social systems should examine opportunities as well as risks. "Co-benefits" is language borrowed from economics, but economic co-benefits are not the focus of this paper. Rather, this paper looks at social co-benefits for energy, food and water security, and land use. Energy, food and water, and land are three broad categories in which to focus assessment of co-benefits; there are probably other lenses through which co-benefits could be seen, such as health. It is important to note again that at this point in time, with seven billion people, the problems we have with energy, food and water security, and land access are questions of availability as well as distribution. As this paper is interested in social justice issues, the germane question is, does this SRM technology allow itself to be deployed in a way that improves the availability and distribution of these resources?

If one thinks of SRM as being a solution to a solar radiation problem, it may not be obvious at first what SRM even has to do with energy, food and water security, or land access. Yet solar radiation affects, and is also affected by, land use and vegetation change as well as solar and biofuel energy production (as modifying the earth's surface changes the albedo, and changing the balance of incoming radiation also affects vegetation). Hence, solar radiation affects agriculture and food security, but agriculture affects solar radiation. Here too, there are complex relationships at work with systems that are not just "natural," but deeply social.

Russell et al. have written about potential co-benefits of climate remediation broadly, noting that co-benefits depend on perspective and are ambiguous. "They point out, or example, 'the purposeful enhancement of net primary production by ocean fertilization can potentially add more carbon to the base of food webs, which could be considered a positive outcome or an unwanted ecosystem disturbance'" (de Baar et al., as cited in Russell, 2011, p. 8). Another real concern with the idea of assessing co-benefits is that these co-benefits could serve as a selling point on controversial technologies. Leach et al., writing about the non-SRM technology of biochar, point out that pro-biochar narratives for promoting biochar-producing cook stoves use the idea of the quadruple win in co-benefits (climate change mitigation, energy, agricultural income, and improved health from reduced cooking smoke) (2011, p. 13). Yet these co-benefits only appear when done in appropriate small-scale interventions. The authors point out that even the promise of small, pro-poor biochar projects are disputed by cynical critics, who argue that "these are merely the thin end of a wedge that will lead eventually to large-scale carbon grabs" (2011, p. 14).

Certainly, critical thinking should be applied to all claims of co-benefits for both SRM and carbon dioxide removal (CDR) geoengineering methods. Terrestrial biological strategies may have an edge when it comes to co-benefits: as suggested by an American Physical Society report on direct air capture of CO_2, "co-benefits like enhancing the habitats of wildlife and improving soil productivity are likely to lead to the pursuit of terrestrial biological strategies well ahead of DAC [direct air capture]—even aside from considerations of cost" (2011, p. 20). A recent report about Negative Emissions Technologies (NETs, which are basically CDR technologies) points out that "several techniques are considered to be worth pursuing at a basic level as 'no-regrets' strategies," including "soil carbon management in organic agriculture, wetland restoration, timber use in construction, afforestation, and if the technique stands up to more detailed scrutiny, regenerative grazing" (McLaren, 2011, p. 61). The co-benefits from biological carbon dioxide removal techniques are more obvious, and in many cases may weigh comprehensive assessments in favor of this family of technologies. But perhaps there are potential co-benefits from SRM schemes as well. It is worth investigating, because if the prospects for co-benefits are very low, that is also important to know.

Let us consider a few examples of SRM schemes that, if deployed in certain ways, might have co-benefits. Ridgwell et al. have proposed a "biogeoengineering" approach to changing the albedo where crop varieties with glossier leaves and/or more reflective canopy morphological traits are planted (2009, p. 146). According to their model, this would have relatively small global effects but greater regional cooling; it might also delay Arctic sea-ice retreat. An advantage of this scheme is "the infrastructure required to create and propagate specific physiological leaf and canopy traits to large-scale cultivation is already in place" (Ridgwell et al., 2009, p. 148). One imagines this idea could play out in various ways. It could be done by large biotech companies that design and patent super-reflective terminator seeds, which creates a lock-in bind, forcing farmers to purchase seeds from that company for the foreseeable future. Or it could be done using a public research program where people own the property rights to the genetic information and subsidized seeds as an incentive. It could be a multilateral, UN-coordinated project, or selectively implemented by national governments in the areas where it is predicted to be effective (the EU, Russia, North America). There is a notable amount of flexibility as to how this specific SRM technology could be developed and used. It could offer future generations substantial choices about how to develop, deploy, or stop using it, though if developed by for-profit corporations, it would reduce choice. The potential co-benefits here could be in the realms of socially just agricultural reform and food production.

In another example, Seitz has proposed using micron-sized bubbles to brighten water, hence increasing solar reflectivity. This has possible co-benefits: dilute hydrosols may reduce solar-driven evaporation from canals (Seitz, 2011, p. 376). Much more research is needed, but this could help with local water conservation in areas facing water scarcity. A further SRM technique, modifying the urban albedo by whitening roofs and pavements, has the co-benefit of energy savings and reduced ground level ozone (Swart et al. 2009,

p. 152). Though models show that global effects would be insignificant (Irvine et al., 2011), urban albedo modifications reduce the urban heat-island effect, a desirable benefit even beyond the global warming context. The modeling by Irvine et al. uses a moderate estimate that roofing and paving modifications in urban areas could increase the reflectivity of incoming sunlight by 10 percent (2011, p. 4). The social structures required to renovate roofs could bring co-benefits in the forms of jobs and urban revitalization, even if this only creates an average global cooling of an estimated 0.11 degrees Celsuis. This would slightly recover Arctic sea ice thickness (Irvine et al., 2011) and could be worth doing for the regional effects and potential co-benefits. It is beyond the scope of this paper to evaluate the existence of social and ecological co-benefits of all SRM schemes; this paper merely argues that it would be a valuable exercise to do so.

CONCLUSION: TRANSFORMATIVE DEPLOYMENT

If we combine evaluating geoengineering ideas based upon selecting a target vision for a possible future, maximizing the choice power of future generations, and identifying possible co-benefits, this may clear the way for a more ethical version of SRM. Assessing technologies on these criteria could offer a framework for evaluation that forces us to be creative about the future we want. We have only talked about content here; the process of doing the assessment is of course crucial. So is what is done with the assessment afterwards. As Lohmann points out, "Just as a technology is never 'just' a neutral piece of machinery which can be smoothly slotted into place to solve the same problem in any social circumstance, so the success of a social or environmental impact assessment is dependent on how it will be used and carried out in a local context" (2006, p. 279). History is probably littered with assessments that have languished in filing cabinets or the digital cloud. It is all too easy to imagine assessment as an exercise in reassurance, a powerpoint sentiment. Or, as Hajer points out, a performance:

> "Cost benefit analysis" with its characteristic "trust in numbers" (Porter, 1995), the variety of "impact assessments" and the way they are employed in the policy process, the necessity of reductionism and indeed the creation of "authoritative data," are particular performances, creating a style of operating which can be analysed as a practice of knowledge production, and which is also a way to try and control an ambivalent situation. (2006, p. 50)

This quote brings us closer to the heart of the politics of knowledge production. Assessments that focus on impacts, cost-benefits, etc., are in fact performing a certain politics. Given that these are status quo methods of operating, and given our legacy of the natural/social binary, it seems improbable that we would be able to include social opportunities and the power of choice in how we consider whether (and how) to use these technologies.

Yet if we appraise these technologies well, we really have a chance at achieving something: whether it is salvaging bright strands of the modernist dream, which told us we could rationally plan and choose to enact better conditions for our species and the world—or some other dream which we invent. The stakes may be high, not just ecologically but also culturally. To fail at addressing climate change ethically means admitting that we have given up on many of the values that shaped our society. An ethical emergency intervention may be the best hope we have for reinventing modernity, for illuminating the gravity of our situation, and for nudging us toward overcoming organized irresponsibility.

Solar radiation management prompts us to consider the future we want. Ashley Mercer and David Keith write from an engineering perspective that clients need to state what they want engineers to design for them: "As we move to explore SRM, we must recognize that the fundamental shift from scientific to engineering understanding requires client inputs" (2010, p. 8). That is, we are all clients of engineers now, and we need to decide what kind of earth we want. This will undoubtedly strike some readers as grim or unfair. Who of us asked to be a client? Yet when we eat the bounty of industrial agriculture, warm our homes, or use goods that are shipped, we are all unwitting clients of someone's designed systems. Perhaps the way forward is to demand a stake in earth systems design and demand that the engineers execute the visions we create. Assessment of social choice opportunities and co-benefits is one tool that we could use to determine our visions.

Ethics of Framing and Rhetoric

Chapter Nine

Insurance Policy or Technological Fix? The Ethical Implications of Framing Solar Radiation Management

Dane Scott

Climate engineering is an incredible, outlandish idea. Yet, investigating how to control the earth's climate system is quickly moving from being considered irresponsible to becoming a serious research agenda. The threats of dangerous climate change are looking grave, and political processes appear ineffectual. This is leading many people to see climate engineering research in a new light. For example, the Intergovernmental Panel on Climate Change (IPCC) recently decided to include a section on climate engineering in its Fifth Assessment Report, to be published in 2013 and 2014. There is a shift in thinking. Climate engineering research has multiplied over the last few years, and that trend is likely to continue. However, climate engineering is obviously controversial. Along with a growth in research will be a heated international debate. Already, in response to growing interest in geoengineering, in 2010 the United Nations Framework Convention on Biological Diversity agreed on a controversial *de facto* moratorium on climate engineering field experiments. It is all but certain that when the IPCC assessment report on climate engineering appears in 2013 or 2014 that debate will "move to center stage" (Hamilton, 2011, p. 1).

Scientists have speculated on a wide array of climate engineering proposals that fall into two broad sets of strategies. One set of strategies aims at large-scale carbon dioxide removal (CDR) from the atmosphere to reduce the greenhouse effect. The other set of strategies has been called solar radiation management (SRM). This general approach aims to cool the earth by reflecting back incoming solar radiation either in the upper atmosphere or at or near the earth's surface. The most discussed SRM proposal involves injecting sulfate aerosols into the stratosphere to reduce incoming solar radiation. The focus of this chapter is on these SRM strategies.

The debate over climate engineering, and SRM in particular, is a high stakes debate. David Keith et al. characterize this technology as having three essential features: it is "cheap, fast and imperfect" (Keith, et al., 2010, p. 426). Mimicking the cooling effect of large volcanic eruptions will cost relatively little money, be highly effective in cooling the planet, but the exact consequences are not known. This combination of features, when added to the threats of dangerous climate change and the deadlocked political processes, makes for a high stakes debate. Unfortunately, if one looks back at the last three decades of science debates over climate change and other issues such as biotechnology, one finds a highly polarized discourse. One would expect the climate engineering debate to be similarly polarized, and all too often characterized by misrepresentations, misinformation, personal attacks, and the like. One clear indication of this is a recent 50-page, glossy publication by the Erosion Technology and Concentration (ETC) Group titled "Geopiracy" (2011). The ETC Group has played a large role in the biotechnology debate, and the "geopiracy" label is making an explicit link to the negative labeling of transgenic crops as a form of "biopiracy."

ETHICS AND FRAMING SCIENCE

There are numerous issues in the areas of ethics, communication, science, and public deliberation that could be researched to shed light on the debate over SRM and climate engineering in general. This chapter will focus on the area of research involving ethics and the framing of science. The communications scholar Matthew Nisbet has published several articles on the framing of science that discuss ethics. He has identified ways in which the framing of science has undermined the ideal of informed, intelligent, democratic deliberations. However, Nisbet cautions that the answer to unethical framing is not to avoid the framing of science. He writes, "Framing is an unavoidable reality of the science communication process. Indeed, it is a mistake to believe that there can be 'unframed' information" (Nisbet, 2009a, p. 1771). One response to unethical framing of science in public debates is for scientists, journalists, and citizens to become more cognizant about ethics and the framing of science in a democratic ethos.

Frames are used to simplify complex issues by focusing attention on some aspects of the issue while excluding others. They suggest what a controversial situation is about and what should be done. However, by simplifying complex issues through selective emphasis, framing requires value judgments on what features of a controversial issue are important and what aspects can be deemphasized or ignored. In the act of selective emphasis, frames influence how people think about issues and what *ought* to be done to resolve them. In cases where important moral issues affecting people's lives and the environment are at stake, framing has ethical dimensions. The framing of science in a democracy where citizen participation is valued "should be designed to promote dialogue, learning, and social connections that allow citizens to recognize points of agreement while also understanding roots of dissent" (Nisbet, 2009b, p. 1771). The following question is explored in this chapter: Is SRM being framed in ways that would contribute to moving public debate in the direction of the ideal of reasonable democratic deliberations?

The novelty and controversial nature of SRM has created a situation where several competing frames are being used to discuss SRM proposals. In scholarly articles and the in popular press, SRM is being characterized as an insurance policy, Plan B, techno-fix, lesser of two evils, sun block, Pandora's Box, and Promethean science, among others. As indicated, the way SRM schemes are framed has important moral and ethical implications for the ways in which the public will perceive and debate these proposals. These various frames implicitly indicate what these climate engineering proposals are for (e.g., an insurance policy) and suggest how we should react to them (e.g., the lesser of two evils). A recent article on public understanding of SRM notes that "SRM is a new technology, and while public awareness may be growing, pubic opinions are in their formative stage and are sensitive to changes in framing" (Mercer et al., 2011, p. 8). In this paper I will discuss two main ways SRM is currently being framed: "insurance policy/plan B" and "technological fix."

Both of these framings are ethically loaded, and much is at stake in how SRM is framed in public discourse. People will react very differently to SRM if it is characterized as an "insurance policy/plan B" versus "techno-fix." Purchasing an insurance policy or having a plan B is generally seen as prudent behavior whereas using a techno-fix is morally ambiguous or simply wrong. In addition, the insurance policy/plan B and technological fix frames capture important perspectives on SRM. For instance, the 2009, Royal Society Report, "Geoengineering the Climate: Science, Governance and Uncertainty," notes that the there are three distinct "perspectives on the potential role for geoengineering" (p. 45). The three perspectives mentioned are: (1) a means of buying time that was lost in the slow-moving international climate change negotiations; (2) an unethical and risky manipulation of earth systems; and (3) "strictly as an insurance policy against major mitigation failure" (p. 45). As will be seen, the first two of these perspectives are captured by the technological fix frame, and the third perspective is a statement of the insurance policy frame. While this analysis will be selective, it is hoped that it will fulfill the goal of provoking awareness and critical reflection on ethics and the framing of SRM. It is clear from recent research in science communication that the way scientific issues are framed has far reaching implications for public debates and ultimately real world consequences. On the one hand, if SRM is an insurance policy, then it is something one never plans to use. On the other hand, if SRM is a technological fix, it is a possible plan of action.

INSURANCE POLICY

Of the three main perspectives listed in the 2009 Royal Society report, the insurance policy and plan B frames see climate engineering "strictly as an insurance policy against major mitigation failure" (p. 45). This framing is common among scientists and journalists who report on the science. The insurance frame has a definite logic or narrative that justifies SRM research. One of the first scientists to use the insurance frame was Freeman Dyson. While working on a climate and energy program in 1977, Dyson published an article introducing an early CDR proposal. Dyson begins by providing the now standard climate emergency scenario to justify research into climate engineering. He writes: "Suppose that with the rising level of CO_2 we run into an acute ecological disaster. Would it then be possible for us to halt or reverse the rise in CO_2 within a few years by means less drastic than the shutdown of industrial civilization?" (p. 288) Dyson speculates that it "should be possible in the case of a world-wide emergency to plant enough trees and other fast-growing plants to absorb the excess CO_2 and bring the annual increase [of CO_2] to a halt" (p. 288). The purpose of Dyson's article was to provide rough calculations that demonstrated the economic and technical feasibility of such a plan in order to encourage research. Dyson warns that this climate engineering plan would be a short-term emergency response that would buy time for the long-term solution of shifting the world economy away from fossil fuels. He then frames the purpose of climate engineering research: "To have plans ready is a form of *life insurance*, valuable even if the threat of catastrophe never happens" (p. 288). By using the insurance frame, Dyson is making an argument by analogy: just as it is prudent to purchase an insurance policy even if you never plan to use it, it is also prudent to research climate engineering even if we never plan to do it.

For better or worse, the repeated use of a particular framing can influence public debate. Over the last two decades, the insurance frame has been frequently used by scientists advocating the need for research into climate engineering. Discussions about climate engineering started in earnest in the 2000s, and the insurance frame frequently appeared in professional articles and in the popular press. For example, in an important 2002 review article,

"Advanced Technology Paths to Global Climate Stability," the authors chose this framing. The team of distinguished scientists who coauthored the article ended their review by remarking that no review of advanced technology would be complete without a discussion of geoengineering. They wrote: "Our assessment reveals major challenges to stabilizing the fossil fuel greenhouse with energy technology transformations. It is only *prudent* [emphasis added] to pursue climate engineering research as an *insurance policy* [emphasis added] should global warming impacts prove worse than anticipated and other measures fail or prove too costly" (Hoffert et al., 2002, p. 986). However, they warned that "large-scale geophysical interventions are inherently risky and need to be approached with caution" (p. 986). Once again, this framing is an argument by analogy that encourages the prudent course of action of buying life insurance, i.e., conducting research. The authors added an additional caveat that Dyson did not mention: climate engineering is *inherently* risky.

Scientists have frequently repeated the insurance frame in the popular press to help people understand the need for research. One of the coauthors of the above report and a leading expert on geoengineering, Ken Caldeira, uses the insurance frame in an Op-Ed piece in the *New York Times* arguing for SRM research. Caldeira provides the standard narrative as part of the insurance policy frame. He glibly asks: "Which is the more environmentally sensitive thing to do: let the Greenland ice sheet collapse and polar bears become extinct, or throw a little sulfate in the stratosphere?" Caldeira (2007). He then recommends that we "think of [SRM] as an insurance policy, a backup plan for climate change" (Caldeira, 2007). Further, he notes that this insurance policy or plan B would be affordable: SRM would be cheap and easy to accomplish. A second example comes from an article, also in the *New York Times*, that discussed the controversy surrounding the game-changing 2006 article in which Nobel-prize winning scientist Paul Crutzen argued for research into geoengineering. In the *Times'* article, earth systems scientist Ralph Cicerone was asked to comment on Crutzen's controversial proposal. Cicerone is quoted as saying: "Geoengineering is no magic bullet. But if done correctly it will act like an insurance policy if the world one day faces a crisis of overheating, with repercussions like melting icecaps, droughts, famines, rising sea levels and coastal flooding" (Broad, 2006). Examples of scientists and journalists repeating the insurance framing like the ones above are numerous. In general they follow Dyson's logic: just as it is prudent to buy insurance, it is prudent to research SRM. Most recent articles are careful to add that SRM should only be a short-term emergency response, and that it is inherently risky.

It is important to discuss a second aspect of the insurance frame. Interestingly, though the insurance frame provides an analogy to argue for research, it also provides one of the main arguments that has been employed against research. The Royal Society's report on geoengineering notes that there is an explicit connection between the insurance frame and "one of the main ethical objections to geoengineering," the moral hazard argument (Royal Society, 2009, p. 39). It seems unlikely that people would be discussing SRM creating a "moral hazard" if SRM had not first been framed as an insurance policy. The Report goes on to explain that the term *moral hazard* is "derived from insurance" and refers to a situation where a newly-insured person takes greater risks because compensation is available (p. 37). "In the context of geoengineering, the risk is that major efforts in climate engineering may lead to a reduction of effort in mitigation and/or adaptation because of a premature conviction that climate engineering has provided 'insurance' against climate change" (p. 37). It may be unfortunate the insurance framing has been extended to frame a main ethical objection to climate engineering. Many scholars working in economics consider "moral hazard" as a value neutral term used to describe behaviors of the insured that will increase losses to private and social insurance systems (Dembe & Boden, 2000, p. 258). In fact, the concept of a moral hazard does not inherently indicate immoral behaviors (Hale, 2009, p. 1).

The moral hazard argument has played an important role in the early discussions about ethics and SRM. Unfortunately, this concept seems to have misdirected discussions on the ethical issues arising from climate engineering. Ben Hale argues in this volume that the "moral hazard arguments against climate engineering fail on their face. They fail not because they are wrong or incorrect, but because they are far too complicated and multilayered to do the work that they are assumed to do" (Hale, this volume, p. 115). It is Hale's view that "geoengineering-related moral hazards are better addressed *more directly* [emphasis added] with other arguments" (this volume, p. 115). The Royal Society Report seems to sense problems with the discussion over the moral hazard arguments as it calls for additional research into these concerns. The report notes that "there is as yet little empirical evidence on whether the prospect of climate intervention galvanizes or undermines efforts to reduce emissions" (Royal Society, 2009, p. 45). The report suggests that "the moral hazard argument requires further investigation to establish how important an issue this should be for decision makers" (p. 45). There are clearly real moral concerns about how people and institutions will interact with SRM research. However, the moral hazard concept from insurance is perhaps not the best place to explore these concerns. The fruitlessness of discussions on the moral hazard argument should raise questions about the aptness of the insurance frame. It is interesting to speculate on the power of frames to direct—or misdirect—discussions about ethics and science policy. The moral hazard discussion may ultimately prove to be a red herring, and it is possible we would never have engaged in this discussion if it were not for the insurance frame.

The analogy between SRM and an insurance policy seems to be extremely limited. These limitations are colorfully alluded to in a 2009 article by Graeme Wood that appeared in the *The Atlantic*. Wood writes, "A pessimist might judge geoengineering so risky that the cure would be worse than the disease. But a sober optimist might see it as the biggest and most terrifying insurance policy humanity could buy—one that pays out so meagerly, and in such foul currency, that we better ensure that we never need it" (Wood, 2009). People act prudently when they buy insurance policies that are trustworthy because they are legally binding agreements and they are confident in the currency of compensation. It is not prudent to purchase an insurance policy if the "compensation" it offers is "meager" and "foul." The analogy between insurance policies and SRM quickly breaks down when one starts looking at many other important features of SRM that citizens need to discuss.

If one was not concerned about promoting climate engineering research, as Dyson and Caldeira were, for example, I do not think the insurance or plan B analogies would jump to mind. SRM is seen by the scientists mentioned above as a short-term, emergency response to a global environmental crisis. Further, they agree that SRM would be inherently risky and impossible to fully test prior to implementation. Given these features, the classes of analogies that jump to mind are safety technologies or emergency medical devices. In fact, if one was looking to replace the climate engineering and SRM frames, perhaps something like "planetary emergency device," might be a good fit. Although that absurd term will never catch on, it does focus attention on the technological analogies that could be a better fit than the insurance or plan B framings.

Numerous technological devices are designed to be used in emergency situations. One possible analogy might be the transceivers used by backcountry skiers. This safety technology's ability to reduce risk is not especially comforting. If a skier is buried in an avalanche, the signal from the beacon allows fellow skiers to locate the buried skier and dig him out. However, time is short, as the buried skier can only breathe for a limited period of time. Like SRM, this is clearly not a reassuring safety technology. If this is "plan B," it should make backcountry skiers want to avoid the risk of avalanches at all costs. However, unlike SRM, avalanche transceivers do not directly create risks to the skier. Perhaps better analogies can be found in emergency medical technologies. One could look for technologies designed to keep a patient alive in an emergency situation but at the same time pose a risk to the patient's health. One analogy might be dialysis, in the setting of kidney failure due to poorly controlled diabetes or hypertension. Dialysis will keep the patient alive but can potentially harm the patient by increasing the chance of infection and other complications. However, these medical technology framings are also limited. As the historian of technology James Fleming notes, individuals can elect to have a medical treatment or not, whereas getting consent from everyone affected by a climate engineering project would not be possible (Fleming, 2006, p. 23). So while it fails to capture this crucial ethical component, it would make a difference if SRM were thought of as being more like a dialysis machine than an insurance policy. Even though both analogies might be used to argue for research, no one would see dialysis as an insurance policy against complacency to take care of conditions such as diabetes or hypertension.

The point of the above exercise is to emphasize the need to look for frames that elicit the relevant discussions. As Nisbet points out, "[frames] endow certain dimensions of a complex topic with greater relevance than the same dimensions might appear to have under an alternative frame" (Nisbet, 2009b, p. 44). Further, "frames simplify complex issues by lending greater weight to certain considerations and arguments over others. In the process, they help communicate why an issue might be a problem . . . and what should be done" (p. 44). The insurance frame emphasizes the function of SRM as a security device and is too narrowly focused on one aspect of SRM to open important areas of discussion. The *technological fix* frame has the advantage of being more direct. This framing might better focus attention on the relevant issues and open discussion.

TECHNOLOGICAL FIX

As opposed to the insurance and plan B frames that focus attention on the positive role of climate engineering research, the technological fix frame paints climate engineering in a malevolent or suspicious light. However, there are at least two distinct ways that the technological fix framing is used. Recalling the perspectives on climate engineering listed in the 2009 Royal Society report, the technological fix frames see these proposals as either (1) a means of buying time that was lost in the slow-moving international climate change negotiations; or (2) simply an unethical and risky manipulation of earth systems (p. 48).

To avoid confusion between these two uses, I will refer to the unethical and risky manipulation perspective as the "techno-fix" frame and the means of buying time perspective as the "technical fix" frame. To clarify this distinction: on the one hand, authors often use the techno-fix frame to place climate engineering within the larger context of a sweeping philosophical critique of western culture's habitual use of scientific technology to solve problems. This framing is not specific to geoengineering; it is used for a variety of emerging technologies, such as agricultural technology, medical biotechnology, and nanotechnology. On the other hand, the technical fix frame is also used to critique a broad variety of emerging technologies. The latter frame focuses attention on two more practical problems associated with using technologies to address social, political, and environmental problems: the tendency for technologies to generate unintended consequences and the fact that technical fixes do not address the root cause of the problem. The ambivalence of the term technological fix makes it inherently confusing. Nonetheless, exploring the distinctions between techno-fix and technical fix usages will shed light on ethics and the framing of SRM. In what follows, I will provide a brief history of the origin of the term *technological fix* and then discuss and analyze the techno-fix and technical fix framings of SRM.

Alvin Weinberg, the influential director of the Oak Ridge National Laboratory from 1955 to 1973, coined the term *technological fix* at the height of technological optimism in the 1960s (Weinberg, 1969b). He characterized a technological fix as the solution to an intractable social problem that results from reframing it as a social problem to framing it as an engineering puzzle. The idea was to redefine social and political problems as engineering puzzles because engineering problems provide clearer solutions. Weinberg lists the major benefits of the technological fix: (1) technological problems are much simpler than social problems—it is easier to define them and identify solutions; (2) technological problems do not have to deal with the complexity and unpredictability of social systems; (3) they provide policymakers with more options; and (4) they can buy time until the problem can be dealt with on a deeper level. SRM proposals are textbook examples of a technological fix. Most discussions of SRM begin, as did this chapter, by stating that the reason scientists and policymakers are considering climate engineering research is due to the incapacity of political processes to deal with the climate problem. The overwhelming complexity of social and political systems seems to have made the problem of global climate change insoluble. But when the climate problem is reframed as an engineering puzzle involving manipulating stocks of atmospheric carbon (CDR) or the solar budget (SRM), crisp, clean, technical solutions seem to emerge.

In the 1960s, Weinberg was advocating technological fixes as a positive social action. However, the term quickly lost that positive meaning. In the introduction to a 2004 collection of essays focusing on technological fixes, Lisa Rosner writes,

> The term *technological fix* is ubiquitous: it is found everywhere in commentaries on technology. Perhaps that is why the phrase is so hard to define. . . . It has become a dismissive phrase, most often used to describe a quick, cheap fix using inappropriate technology that creates more problems than it solves. (p. 1)

Yet despite the common use of the term, technological fixes occupy a paradoxical status in industrial societies: while the term is used as a derisive label, western cultures by-and-large demonstrate an overriding preference for technological fixes. The preference for solving problems with technology is driven by deep-rooted habits of thinking and tremendous institutional and economic momentum. The fact that "technological fix" has become a common dismissive term in public rhetoric hides its significance as a central concept for understanding the roles technologies play in contemporary life. The following analyses of the philosophical "techno-fix" and pragmatic "technical fix" framings will hopefully demonstrate the concept's importance for framing public deliberations over geoengineering.

The "Techno-Fix"

As noted, of the three main perspectives on climate engineering the philosophical techno-fix framing is associated with the view that sees these technologies as an unethical and risky manipulation of earth systems (Royal Society, 2009, p. 48). Once again, the "techno-fix" framing of climate engineering places this technology within a larger philosophical critique of modern technological society. For example, in the introduction to a new book, *Techno-Fix: Why Technology Won't Save us or the Environment*, the authors' stated goal is to question "the primary paradigm of our age: that advanced technology alone will extradite us from ever-increasing load of social, environmental and economic ills" (Huesemann & Huesemann, 2011, p. xxiii). The movement to challenge the techno-fix paradigm, if you will, began roughly in the middle part of the twentieth century and remains strong among environmental thinkers and activist groups that oppose climate engineering research today. For example, some of the most vocal groups against geoengineering: the ETC Group, Hands Off Mother Earth (H.O.M.E.), and Earth First!, consistently use the techno-fix frame. It will be helpful to take a quick look at development of the key ideas behind the techno-fix frame and its implications for the climate engineering debate.

One the influential people who helped develop the philosophical critique of technology and society was Lynn White. In his article "The Historical Roots of our Ecological Crisis," White sets an agenda by arguing that the source of the twentieth century's ecological crisis is the colossal power created by the union of science and technology in the modern era along with a worldview that justifies the use of that power to control nature. These kinds of dark thoughts about technology and modern culture influenced many thinkers in the second half of the twentieth century. For example, in the 1980s philosopher Alan Drengson wrote an article on technological fixes. In that article he labeled the worldview guiding industrial societies as the "technocratic and instrumentalist" view. He argued that the technocratic and instrumentalist worldview is driving the repeated application of ever more powerful technological fixes, causing escalating environmental problems that pose severe risks to the earth (Drengson, 1984). To stop this destructive pattern, he urged us to abandon the belief that humans have "power as masters and controllers of nature" (Drengson, 1984). In the 1990s, Eric Katz, who has also written articles challenging the use of technological fixes to solve environmental problems, summarized these ideas when he wrote: "The insidious dream of domination can only end by respecting freedom and self-determination, wherever it exists, and by recognizing the true extent of the moral community in the natural world" (Katz, 1992, p. 237).

This way of thinking about history, technology, society, and environmental problems is the reason many environmental thinkers and activists are using the techno-fix frame for SRM. A recent article in *Environment* notes the following: "For groups and individuals who see climate change as a symptom of a social and economic order that is inherently unsustainable, geoengineering represents the worst kind of 'techno-fix'" (Corner & Pidgeon, 2010, p. 31). A full expression of the techno-fix frame is found in a recent article by the philosopher Clive Hamilton. He notes that western societies are unwilling to address the "deeper questions of the roots of the climate crisis" (Hamilton, 2011, p. 17). We are avoiding the uncomfortable conclusions "about the social dysfunctions and the need to directly challenge powerful interests. In this way, global warming becomes no longer a profound threat to our future security and survival but just another problem that must be approached like others. Calls for a *techno-fix*, including geoengineering, are thus deeply conformable with existing structures of power and a society based on continued consumerism" (Hamilton, 2011a, p. 17). For thinkers like White, Drengson, Katz, and Hamilton, solutions to environmental problems like climate change could never be powerful technologies like SRM. Western societies must challenge their fundamental values and worldview to find solutions to climate change.

There is something to these calls to rethink the philosophy guiding modern societies, but the techno-fix framing of climate engineering is clearly polarizing. It is a catchphrase that is designed to immediately prejudice one against SRM. Because of this, it closes off discussion between groups with contrasting positions while reinforcing biases of members of like-minded groups. Speaking on ethics and the framing of science, Nisbet and Scheufele (2009) write: "Framing should be used to design communication contexts that promote dialogue, learning, and social connections and that allow citizens to recognize points of agreement while also understanding the roots of dissent" (p. 1771). Because the techno-fix frame is connected to a sweeping condemnation of western technological culture, it leaves little space for promoting dialogue between opposing groups. Moreover, the philosophical techno-fix framing often uses an "us versus them" strategy.

The people interested in promoting climate engineering research are sometimes referred to as "techno-fixers." They are seen as promoting anti-ecological worldviews, and are associated with illegitimate economic powers. This kind of polarizing discourse is demonstrated in an editorial by Clive Hamilton (2011b) titled, "The Clique that Is Trying to Frame the Global Geoengineering Debate," which appeared in *The Guardian*. He refers to an unidentified group of North American scientists who have been consistently involved in discussions about geoengineering as the "geo-clique." He goes on to say: "The Promethean dreams of the geo-clique [are] perhaps expressed most starkly by its sometime mentor, Pentagon 'weaponeer' Lowell Wood, when he declared, 'We've engineered every other environment we live in, why not the planet?'" Hamilton contrasts the Promethean dreams of the geo-clique with the more humble view in Europe, which accords the earth much respect. In Europe, he writes, there is a "historical reservoir of mistrust for the good intentions of humans intoxicated with technological power" (2011b).

The sweeping philosophical nature of the techno-fix frame is likely to reinforce polarized opinions and lead to ideological conflicts. It is important, even paramount, for western societies to have serious philosophical discussions about society, technology, and the environment. The radical nature of climate engineering should open up a space for that discussion. It seems unlikely that the techno-fix framing of climate engineering will promote that dialogue. In terms of framing climate engineering in a way that invites reasonable dialogue, the techno-fix framing seems far from helpful. The technical fix framing, however, is a more promising frame to initiate these kinds of discussions.

The "Technical Fix"

So far it has been argued that the insurance policy/plan B frame misdirects and limits discussions and that the techno-fix frame is likely to short-circuit discussions on SRM and polarize the debate. It is unfortunate that the dual uses of the technological fix frame creates confusion, as what I am labeling for the purposes of this discussion "the pragmatic technical fix" framing has the potential to point discussions in the right direction.

To recall, of the three distinct perspectives on the "potential role for geoengineering," mentioned in the 2009 Royal Society Report, the technical fix frame represents the buying time that was lost in the slow-moving international climate change negotiations perspective (p. 45). One example of a scientist using the pragmatic technical fix frame is found in a 2010 article, "Engineering the Planet." In this article, David Keith writes, "Geoengineering entails the application of a countervailing measure, one that uses an additional technology to counteract unwanted side effects without eliminating their root cause, a 'technical fix'" (p. 494). He goes on to argue that SRM might be legitimately used as a temporary, technical fix to augment slow-starting mitigation efforts to avoid severe climate risks during peak periods of GHG concentrations. Stated differently, SRM might provide policymakers with an additional tool that could be used to buy time while long-term mitigation efforts take effect.

The framing of SRM as a pragmatic technical fix corresponds with Weinberg's list of benefits of technological fixes: it provides policy makers with more options and can buy time until a deeper solution is found. However, as was seen, technological fix has lost most of the positive meaning Weinberg hoped it would have. Nevertheless, many developed societies seem to have an overriding preference for technological fixes, and it is good that this frame creates suspicion. The technical fix frame causes people to be suspicious of SRM while immediately drawing attention to two familiar defects associated with using technologies to solve social and environmental problems. Because of our experiences with technologies over the last century we know they tend to interact with cultural and natural systems to generate unintended consequences: that they are often merely expedient solutions that do not address the underlying problem. Yet we still often use technical fixes and they are sometimes helpful in addressing problems. Timothy LeCain points out that "data would suggest that, contrary to popular perceptions, environmental technical fixes have indeed solved many environmental problems" (LeCain, 2004, p. 177). However, this is true only in a restricted sense; a major characteristic of technological fixes is that they are ambiguous. The judgment that they have provided a "solution" depends on who is defining the criteria for success and how they are defined. If we are to consider researching SRM as a pragmatic technical fix to buy time, then the focus of deliberations on SRM should be on these issues: What are the possible *unintended consequences* of SRM and how would SRM affect the "root causes" that ultimately must be addressed?

This point is reinforced by Corner and Pidgeon's survey of the social and ethical issues of climate engineering (2010). They note that "concerns about whether scientists and engineers have the capacity to safely mitigate the unintended technical and environmental consequences of geoengineering will play a central role in the geoengineering debate" (p. 31). However, discussions about the safety of SRM must be about more than utilitarian calculations of risks versus benefits. Risk/benefit calculations do not get at the deep worries over technical fixes like SRM. The 1996 book by Edward

Tenner *Why Things Bite Back: Technology and the Revenge of Unintended Consequences* sheds light on the deeper fears. In that book he makes an important distinction between "side effects" and "revenge effects." He notes: "a revenge effect is not the same thing as a side effect. If a cancer chemotherapy treatment causes baldness, that is not a revenge effect; but if it induces another, equally lethal cancer, that is a revenge effect" (p. 7). Revenge effects are not trade-offs like side effects, which represent the "costs" we might weigh against the "benefits" of SRM. For example, most cancer patients are willing to pay the costs of baldness when compared to the benefits of chemotherapy. However, revenge effects are unforeseen consequences of technologies that are worse than, or at least as bad as, the original problem. One much discussed recent example of a revenge effect of a safety technology is football helmets. Football helmets are designed to protect the brain from injury, yet they may be having the revenge effect of contributing to the tragedy of early onset dementia and Alzheimer's. This safety technology may be giving players a false sense of security leading them to abuse their brains with repeated blows to the head. Looking at an SRM scheme, putting sulfate particles in the upper atmosphere has the potential cost or side effect of the whitening of the sky (Robock, 2008, p. 16). However, many possible revenge effects from using SRM have also been raised. These potential revenge effects and their possible effects on long-term solutions should be the main focus of research and discussion about possibly using SRM as a pragmatic technical fix for buying time while climate change is dealt with at a deeper level. To develop this point, I will briefly mention a few examples of possible revenge effects that are starting to be discussed.

In an insightful statement, Tenner writes, "the danger of technology is precisely [the] feeling of immunity from natural events" (p. 229). Tenner lists the avalanche beacons mentioned earlier as a technology that has created this kind of revenge effect. He writes: "The more reliable the control of avalanches is thought to be, the bolder some skiers and other winter athletes will feel" (p. 229). There are numerous examples where safety devices have interacted with human behavior and cultural systems to create revenge effects. Safety technologies can cause people to take greater risks, and they can be misapplied, leading to greater harms. In the case of SRM as a pragmatic technical fix for buying time, one revenge effect scenario that should be discussed is where key political leaders decide to delay or take less aggressive mitigation efforts because SRM may be available. Samuel Thernston (2010), writing in *The American: The Journal of the American Enterprise Institute*, argues for research into SRM as a way of buying time to allow for a slow transition from a fossil fuel economy. He writes: "Seen in the proper light, geoengineering is potentially the key to unlock the mitigation puzzle—a way of controlling climate risks during the many decades that it will take to transform the global energy system." Once again, if SRM is seen as a safety technology for buying time, it could cause political leaders to take greater risk with climate change by being less aggressive at mitigation and adaptation efforts. Recalling the earlier discussion on the moral hazard frame, this kind of revenge effect is perhaps what the "moral hazard" frame from insurance is trying to raise. However, as was noted, Hale points out severe limitations with this concept as applied to SRM and that more direct arguments could be found elsewhere. Also, it was suggested that the relationship between safety technologies and risk-taking might be a more direct way of exploring the concerns poorly raised by the moral hazard argument. It is worth noting that there is nothing necessarily immoral about taking greater risks because of safety technologies. The focus of the discussions, then, should be on political leaders' and policymakers' motivations for taking risks—rather than the mysterious phenomenon of the moral hazard. More specifically, if political leaders, who hold the public's trust, were motivated by a vested interest to use SRM as a reason for less aggressive mitigation policies that could be a serious ethical breach. If mitigation efforts were delayed and SRM deployment failed to work as designed, this could create a severe revenge effect. This is particularly worrisome in the case of SRM. Many safety technologies are tested and refined as people use them, and we often learn from tragic experiences. However, SRM cannot fully be tested until it is actually deployed. The implications of an SRM deployment not working as designed are cause for serious concern given the unprecedented scale of this technology. In addition, safety technologies fail for many reasons beyond design problems or how they interact with complex natural systems. For example, it is easy to imagine SRM not being deployed correctly for numerous political reasons.

One of the most important questions the pragmatic technical fix framing should cause people to ask is: What is the underlying problem of the failure to act on the threat of dangerous climate change? It might be argued that one of the underlying issues that must be addressed to find a long-term solution to climate change is inadequate cooperation among nations. In fact, one of the justifications for using SRM is to buy time to overcome the difficulties nations are having building the necessary cooperation for long-term agreements to cut GHG emissions. However, SRM could have the revenge effect of making the problem of cooperation worse, not better. Because of their divergent interests, it has proven to be especially difficult for the major emitting countries, particularly China and the United States, to cooperate on international climate change agreements. A major emitting country or group of countries might decide that the dangers of climate change had become sufficiently severe that it was in the country's best interest to act unilaterally and deploy an SRM project. This act would surely be seen as a defection from international climate change negotiations and make the cooperation required for mitigation efforts much more difficult. It is not impossible to imagine, for example, that though China's and the United States' interests have diverged on climate change negotiations, they might see their interests converging on an SRM project. While this scenario seems improbable at the moment, it would be disastrous for international negotiations. In a 2009 article, Victor et al. note the potential for unilateral climate engineering could create strong incentives for some countries to defect from cooperative efforts in order to avoid the high cost of emissions controls. Again, this kind of revenge effect would need to be anticipated and become the object of much discussion.

Another underlying problem—which SRM might affect—that will need to be addressed in order to find a long-term solution to the climate problem is the lack of citizen engagement. If political leaders in democratic societies are to make the difficult and politically risky choices to address climate change they will need popular support for these efforts. However, in key countries like the United States, the citizens are not, by and large, engaged in this issue. While U.S. citizens are concerned about climate change, it remains a low priority. Anthony Leiserowitz reported in a 2007/2008 survey that "environmental issues are consistently viewed as relatively low priorities compared to other national issues and that global warming remains a relatively low priority among other environmental issues" (p. 29). In the survey, climate change was ranked tenth out of ten national issues. Could using SRM buy time while people get engaged in this issue, or could it have a revenge effect and further disengage citizens?

Albert Borgmann, the influential philosopher of technology, sheds light on how technologies can cause us to be disengaged from others and the world around us. Though Borgmann's analysis focuses on technologies people use every day, it might be applied to SRM. One of Borgmann's important insights is captured in his notion of the device paradigm (Borgmann, 1984). An essential feature of modern technologies is they tend to make things available for us without imposing burdens. One consequence of disburdening us from various tasks is devices lessen our engagement in the world. Many of these burdens we are happy to offload to machines. However, in an interview Borgmann remarks the task of philosophers of technology is "to point out what happens when technology moves beyond lifting genuine burdens and starts freeing us of burdens that we should not want to be rid of" (Wood, 2003, p. 22). In looking at SRM, then, one question to ask is the following: Is this particular technology lifting burdens from us that we should not want to be rid of? Clearly, one of the underlying problems with climate change policy is the fact that the public is not sufficiently engaged in the issue to create the political will for action. Too few people are seriously engaged in what looks to be one of the great moral issues of our age. According to Borgmann, a large part of the reason for this lack of engagement in the climate change issue is that citizens, at least in the United States, are immersed in a "paradigm of instantaneous, ubiquitous, and attractive availability" of modern technologies (Borgmann, 2011, p. 176). It would be a clear mistake to pursue SRM research to create a "device" that would further disburden citizens from taking moral responsibility for the climate issue. The twin problems of building international cooperation for effective climate change treaties and transitioning economies away from carbon energy sources will take highly engaged and motivated citizenries. It is true that SRM will be a highly contentious political issue that will engage some citizens. However, this kind of technology would call for a highly bureaucratic and centralized control that would involve relatively few experts to implement, monitor, and maintain. If SRM follows the device paradigm, it is likely to have the revenge effect of further disengaging many citizens from this issue, undermining support for difficult political decisions and hard economic choices.

CONCLUSION

The above brief examples of possible revenge effects are listed to demonstrate how the technological fix frame might focus discussions on the relevant issues. However, the ambiguity between its sweeping philosophical use, "techno-fix," and more pragmatic use, "technical fix," may ultimately cause this framing to create too much confusion. Nonetheless, I do believe that it points to the kinds of issues that scientists, policymakers, and citizens should be discussing: potential unintended consequences/revenge effects and the nature of the underlying problems SRM is failing to address and might make worse.

In summary, SRM is a confusing and novel technology, and "while public awareness may be growing, pubic opinions are in their formative stage and are sensitive to changes in framing" (Mercer, et al. 2011, p. 8). Framing of SRM cannot be avoided. As research moves forward and international debate over SRM and other climate engineering schemes heats up, there is a great need to pay attention to how these technologies are framed. Nisbet and Scheufele (2009) provide some rough guidelines for framing science:

> Framing can be used ethically by prioritizing dialogue and bottom-up citizen expression, by avoiding false spin or hype and remaining true to what is conventionally known about a scientific topic, and by avoiding denigration of social groups and the advancement of partisan causes. (p. 1771)

How SRM and climate engineering are framed by scientists and journalists will influence the quality of public deliberations. The last few decades of the climate change debate should be a cautionary tale on the severe costs of promoting an ideologically polarized debate that is focused on the wrong issues. As was noted earlier, one response to unethical and unfruitful framing of science in public debates is for scientists, journalists, and citizens to become more cognizant about ethics and framing of science in a democratic ethos. Hopefully this chapter has provoked interest and reflection on these important issues.

Chapter Ten

Public Concerns about the Ethics of Solar Radiation Management

Wylie Carr, Ashley Mercer, and Clare Palmer

"Talk straight, make sure, don't screw up, and don't forget anything."

—Respondent 747

In the past five years, a shower of reports published by think tanks and scientific bodies have discussed future prospects for the geoengineering technologies currently being proposed in light of rising greenhouse gas (GHG) emissions. These reports have characteristically emphasized that social and ethical considerations should play a key role in geoengineering research and have insisted that public engagement is critical to the process of developing and debating geoengineering technologies. For example, the influential 2009 Royal Society report "Geoengineering the Climate" stated, "The greatest challenges to the successful deployment of geoengineering may be the social, ethical, legal and political issues associated with governance, rather than the scientific and technical issues" (2009, p. xi). One of the key recommendations of this report was to "initiate a process of dialogue and engagement to explore public and civil society attitudes, concerns and uncertainties about climate change" (2009, p. xii). Likewise, a recent report on geoengineering produced by the Bipartisan Policy Center (BPC) emphasized that geoengineering raises "new ethical, legal, and social issues of broad public concern" (2011, p. 19) and that "public engagement must inform the program agenda" (BPC, 2011, p. 17).

These types of explicit calls for public engagements and ethical inquiries on emerging technologies have grown in recent decades (Rogers-Hayden & Pidgeon, 2007). While both ethics and public engagement are increasingly considered critical and urgent aspects of the development of emerging technologies, they are rarely brought together in a meaningful fashion. Ethicists construct logical arguments, debate these arguments, and come to (often conflicting) ethical conclusions. Social scientists utilize engagement methods such as surveys, interviews, citizen juries, and focus groups to elicit public perceptions that may or may not be incorporated into decision-making processes. But rarely are the arguments of philosophers and the concerns of the public brought into direct contact with one another. So far, work on geoengineering has followed a similar pattern. Despite being pursued concurrently, the small but growing body of academic ethics research on geoengineering (e.g., Gardiner, 2010; Hamilton, 2011a, 2011c; Jamieson, 1996, 2010; Preston, 2011) has remained largely separate from public engagements that probe knowledge about, and opinions of, various forms of geoengineering (Leiserowitz, 2010; Mercer et al., 2011; Godfray, 2010; Spence et al., 2010).

This disciplinary separation represents an opportunity lost. The potential synergies of philosophy and social science are notable—particularly in the context of emerging technologies. Social science research has indicated that public concerns about emerging technologies naturally gravitate toward social and ethical issues (Wilsdon & Willis, 2004; Wynne, 2007). Previous public engagements have revealed areas where average citizens questioned foundational assumptions being made by scientists and/or policymakers and asserted different visions of technologies and their future roles in society (Reynolds & Szerszynski, 2007; Walmsley, 2009). In other words, the basic concerns that individuals have about emerging technologies are inherently, if not always explicitly, linked to ethics. As such, social scientists are increasingly attempting to create space in public engagement processes for the discussion of normative issues (Reardon, 2007).

If these types of processes were brought into contact with philosophical reflection, it seems plausible to expect mutual enrichment. It might be that particular ethical concerns that philosophers have not considered, or have not yet clearly articulated, arise in public consultation. Conversely, philosophers may have persuasive arguments that could reinforce or calm certain public concerns or potentially raise problematic issues in public awareness. Philosophical work can help in interpreting and providing explanations of fundamental ethical concerns—such as concerns about justice or relations between human beings and nature—that may underpin what emerges from public responses. Collaborations of this nature could result in enriched understandings of social and ethical issues related to emerging technologies. In fact Borgmann (2006, p. 15) has described such connections as inherent and unavoidable:

Social science without ethics is aimless; ethics without social science is hollow. In fact, the two fields inevitably overlap. There is no social science research that is not tethered, however indirectly, to concerns of social justice and human flourishing, and there are not ethical reflections that fail to appeal somehow to the actual human condition.

PURPOSE AND METHODS OF THIS PAPER

This chapter aims to initiate an interdisciplinary and mutually beneficial conversation between ethics and social science on geoengineering. Co-written by two social scientists and an ethicist, this paper brings together data from an international survey of public perceptions of geoengineering with current research on the ethical questions about geoengineering, either raised by ethicists directly, or found in the ethics sections of major geoengineering reports. We look for areas where philosophical concerns dovetail with, develop, or reinforce public concerns; where they tend to diverge from such concerns; and where there are worries that seem important either to the public or to philosophers that do not arise in the other group. We suggest that dialogues of this kind could improve both public and philosophical engagement with geoengineering and improve the research and development of solar radiation management technologies.

Some stylistic challenges arose in merging our disciplinary backgrounds. Social scientists typically inhabit a world of quantitative and qualitative social data and tend to prefer empirical statements to normative ones when describing study findings. But we found in constructing this paper that holding fast to disciplinary conventions can inhibit meaningful cross-disciplinary exchange. Presenting survey methods and results as is commonly done in social science may render findings of consequence to an ethics audience uninteresting or inaccessible. As such, our presentation of the data breaks with some social science conventions. We do not attempt to quantify statistically various positions but focus instead on the significance of particular quotations for exposing shared and divergent concerns between the public and philosophers about the ethics of geoengineering.

The Survey

The data presented below come from an international survey of public perceptions of one branch of geoengineering proposals referred to as solar radiation management (SRM). This survey focused on a particular SRM approach that would enhance atmospheric albedo by increasing the concentration of sulfuric acid droplets in the stratosphere (Royal Society, 2009). Scientists hypothesize that this technique could lower global average temperatures relatively quickly and cheaply by reflecting a small percentage of incoming sunlight back into space. This type of intervention presents some significant ecological, social, political, and economic risks and uncertainties at varying scales from the local to the global. Both because of its potential benefits and risks, stratospheric aerosol injection has received significant political and scientific attention in recent years and was the focus of this social science investigation.

The survey was conducted in Canada, the United States, and the United Kingdom in November and December of 2010. The survey questionnaire was designed to assess respondent familiarity with geoengineering more broadly, with SRM in particular, and respondent opinions about these topics. The large majority of survey questions were close-ended, meaning they presented respondents with limited answer choices that were statistically analyzed to make generalized statements about the broader population (see Mercer et al., 2011 for a quantitative analysis of this data). Three survey questions, however, were open-ended, allowing respondents the opportunity to describe their views in their own words. One such question asked respondents, "If you had the chance to talk to Solar Radiation Management researchers and decision makers, what would you say or ask?" Preceded by a short, unbiased description of SRM, to help provide standardized information to all participants, this open-ended question provided a place for respondents to discuss anything they felt relevant to SRM research or policy. The responses to this question highlight the significance of ethical concerns that members of the public have about SRM and provide insights into the nature of those concerns. Based on our analysis, just over 69% of the 2,809 survey respondents who chose to answer this question discussed issues that we identified as pertaining to ethical aspects of SRM. This indicates that the ethics of SRM are extremely salient to members of the public, even when encountering this topic for the first time.

Below we present the ethical concerns of the public using their own words.[1] We have included a large number of quotations to provide readers with an indication of how similar ethical concerns were expressed in different ways. We bring these public concerns together with those expressed in the professional ethics literature, drawing on published material from a number of academic sources. We have classified the quotations related to ethical aspects of SRM into seven broad themes: (1) negative side effects; (2) playing God and messing with nature; (3) the wrong solution; (4) distribution of harms and benefits; (5) the role of science, trust, and public participation; (6) governance of SRM technology; and (7) climate denial. These themes often overlap, and respondents frequently refer to more than one theme in a single quotation. However, there is sufficient distinctiveness between each concern for it to be useful to separate them out.

We hope this approach of coupling social science data with professional ethics literature will allow an ethics audience to see the salience of ethical concerns to the public, to compare the ways in which average citizens describe those concerns to the ways professional ethicists treat similar issues, to see where there are different or competing concerns, and to begin envisioning how future interdisciplinary research, particularly on geoengineering, could benefit from further integration of public engagements and professional ethical inquiry.

ETHICAL CONCERNS ABOUT SRM

1. Negative Side Effects

Many survey respondents raised major concerns about the potential for unintended, negative side effects of SRM. Some respondents raised fairly sophisticated technical questions:

> While I understand that volcanoes release sulfur particulates . . . is the particulate release intended to be on a similar magnitude, or much larger? Will these particulates be prone to causing an increase in lung cancer/respiratory disorders? Is it possible to increase the earth's albedo to a point where we could cool the earth from the surface, rather than atmospherically?

However, technical questions were rare in comparison to general concerns about the potential impacts of SRM on humans, plants, animals, and ecosystems:

> Would these solar emissions that you shoot into the sky have any impact on the health of humans, animals or plant life and will it affect the air we breathe?

> We were told that smoking was not harmful to our health. We were told that asbestos was not harmful to our health. . . . What kind of health problems might be expected or are we just changing one problem for another perhaps more costly?

These types of comments were more focused on the effects than the effectiveness of SRM proposals. Respondents expressing interest in potential side effects were less interested in the technical feasibility of SRM and much more concerned with the possible negative impacts. Many respondents were particularly concerned about side effects that experts might fail to predict:

> The thing that really worries me is the unforeseen effects. However much research you do, you cannot allow for all possible outcomes and once it's done it can't be undone. You cannot know for certain what the effect will be, however much research is done.
>
> You will never be able to fully determine the effects of Solar Radiation Management on nature. Even if there are no short term effects, there will be long term effects that may cause more damage than initial estimates.
>
> Manipulations of nature by humans have a history of turning out very badly, with unexpected side effects—such as introducing alien species of plants or animals into other areas—and even the effects that we have already had on the climate. Did anyone expect what has happened to happen? And we don't know for certain what changes to climate have been effected by humans and what changes are natural. Doing something on a global scale could be extremely risky.... There could be wild extremes of changes that were not predicted.

These responses embody a widespread sentiment among survey respondents that SRM has the potential to bring about unforeseen, negative consequences. These concerns are about more than *risks* or the quantification of the chances of foreseen consequences coming to fruition: these concerns are about uncertainty and ignorance or about not knowing the relevant parameters of natural systems or how they interact (Shackley et al., 1998; Wynne, 2006). This sentiment was nicely captured in the following response:

> Testing Solar Radiation Management on a small scale over several years or seventy years for that matter would never allow you to predict what the effects would be on a global scale, our planet is unpredictable, we should never assume we know exactly how it works . . . taking risks on our planet based on small scale research is frightening to me and I am sure many others also.

In worrying about unforeseen negative impacts of SRM on humans, the public shares the concerns of ethicists and indeed geoengineering researchers as well. Admittedly, some of the specific risks raised by the public do not appear in the ethics literature. For instance, the literature does not discuss direct human health impacts from the use of sulfate aerosols (presumably because ethicists writing on this subject are familiar with scientific literature on the amounts and atmospheric location of the sulfates that would be involved, which makes such problems unlikely). But interestingly, several of the public responses explicitly refer to the possible impact of SRM on nonhuman nature—on the interests of individual living beings and on the health of ecosystems. Such harmful human impacts on organisms and ecosystems have been featured in environmental ethics (for instance, Rolston, 1989; Singer, 2009; Taylor, 1986) and have to some extent played a role in broader worries about climate change (Nolt, 2011b; Palmer, 2011). But there is very little about these kinds of impacts in the geoengineering literature, though environmental ethicists would be expected to have concerns about this issue. More work to explore the ethical questions raised by SRM in terms of its direct impacts on nonhuman organisms and systems seems called for.[2]

Respondents were not only concerned about the *risks* of particular side effects but also about associated *uncertainty* (where the probability of particular outcomes is not known) and *ignorance* (where some of the possible outcomes remain unknown). Moreover, the comments above suggest that some respondents do not believe that further research will result in all possible effects becoming known. Such sentiments mesh interestingly with Sarewitz's (2004) claim that additional research can actually open up new areas of uncertainty rather than reducing it. This worry about uncertainty and "unknown unknowns" has been a major concern in all the relevant literature. The IPCC (2007d, p. 15) notes "Geo-engineering options . . . remain largely speculative and unproven, and with the risk of unknown side effects." Unsurprisingly, this concern has also featured in the ethical literature. As long ago as 1996, Jamieson argued that "The problem with our technological interventions is that we often don't know the price of the meal in advance or even the currency in which it will be extracted" (p. 327). In the ethics literature more broadly, some form of precautionary approach is often defended in cases where outcomes are uncertain (e.g., Gardiner, 2006a). Precaution is not explicitly mentioned in these public responses but may be a way to conceptualize how such concerns could gain traction in a policy context. It is worth noting, though, that not acting for precautionary reasons can itself generate alternative risks (Sunstein, 2005), as there are uncertain, potentially negative outcomes from *not* using SRM in the absence of GHG mitigation.

In the case of ethical concerns about side effects, philosophers and the public have relatively similar responses to SRM. The communication of better information about the known risks of SRM to the public (for instance, that threats to stratospheric ozone are much more likely than threats to human respiratory health) would bring these concerns closer. But worries about uncertainty and "unknown unknowns" are common to both groups. This overlap suggests one key area where social scientists and ethicists could work together to encourage more transparency about risk, uncertainty, and ignorance associated with SRM technologies, encouraging more open and democratic discussions about both science and policy.

2. Playing God and Messing with Nature

A second ethical concern raised by respondents related to "messing with Nature" or "playing God." This overlaps with concerns about side effects, as some respondents feared that "messing with Nature" could negatively rebound on human beings. But there was also a distinct concern about the ways in which human beings understand themselves and about appropriate human relations with the rest of nature:

> What gives you the right to mess about with nature and cause a potentially worse disaster than you're trying to prevent?

> I would say, Stop playing God, there is no easy way out. We must be logical and not go on making more problems for ourselves—You are making us all feel like goldfish in a blender. Thanks for nothing!

The central concerns here seem to be whether humans have a "right" to "mess about" with the natural world and whether it is appropriate for us to shape the earth's climatic systems in accord with human interests. Here again, public concerns resembled those of ethicists, though ethicists have raised several different issues in this context. One kind of worry is about "the end of nature," as articulated by McKibben (1989) in the context of anthropogenic climate change. It is possible to argue that intentional geoengineering takes further steps toward "ending nature" because nature would no longer be "independently-directed." Keith (in Goodell, 2010, p. 45) relatedly describes geoengineering as potentially bringing about "the end of wildness," whereas Hamilton (2011c, p. 21) sees SRM as "a conscious attempt to overcome resistance of the natural world to human domination, the last great stride toward total ascendancy." Something like these arguments appears to be lurking in the public responses outlined above.

A second, related issue that also appears in these responses concerns human self-regard and the idea that intentionally controlling the earth's climate is an example of human hubris, arrogance, or self-aggrandizement. Hill (2007, p. 684) calls attitudes like this a "failure to appreciate one's place in nature" and a lack of "proper humility." As Hamilton (2011a, p. 14, quoting Coady) puts it, one could see SRM as "an unjustified confidence in knowledge, power, and virtue beyond what can reasonably be allowed to human beings." Although the respondents above might maintain that the problem with such overconfident attitudes is that they are *mistaken*—we *do not* in fact have the control over the climate we imagine—the attitudinal worry could remain even if humans *were* in (effective) control of the climate. That we *can* do something does not mean that a good or virtuous society *should* do it; as the second respondent above suggests, this might be taking "an easy way out" rather than addressing a more fundamental problem. Indeed, this may reflect another attitudinal issue—of human laziness and, perhaps, of superficiality. Some environmental ethicists (e.g., Sandler, 2007) have developed accounts of environmental virtues, and Gardiner has specifically, though briefly, discussed these in the context of SRM (Gardiner, 2010, p. 303). The idea that geoengineering is a kind of human over-reaching that reflects negatively on us (or, rather, on *some* of us) arises so widely in public responses that further philosophical work on "playing God" and human virtue (including the idea of collective virtues) seems called for.

Preston (2011) has recently raised the question whether, in certain circumstances, deployment of SRM could be morally permissible despite these concerns, suggesting that such worries should not necessarily be taken as *decisive*. After all, the attitudes displayed by allowing increasing GHG emissions are *also* morally troubling. With continued emissions, there could at some time in the future be a threat of runaway climate change, such that the risks of SRM appear to be much the lesser evil (Preston 2011, p. 471). Such arguments—though not made by any respondents here—are important and indicate a potential need for further public discussion to explore ethical arguments about how, when, and whether justifications for the use of SRM may ever override deep ethical misgivings about intentionally manipulating the global climate. This may be a case where an important philosophical argument such as Preston's could form the foundation of future public engagement processes.

3. The Wrong Solution, Better Alternatives, and Technological Fixes

A third concern was that SRM is the *wrong solution* to the problem of climate change—a "technological fix" rather than something that gets to the heart of the problem. As such, a number of respondents asked whether alternatives to SRM might exist:

> Would there be any other alternative to Solar Radiation Management or other technologies similar that would solve global warming?

> What are the other options? Do we have better solutions?

> Are there any other alternatives currently being studied?

Some participants even suggested alternatives that they felt might be worth pursuing in place of SRM:

> Your time would be better spent trying to develop technologies that can remove substantial amounts of the emissions humans have emitted.

While this individual refers to technologies that *are* currently being researched and developed, namely, carbon dioxide removal technologies, the overall message is that some members of the public were skeptical that all options for addressing climate change were being explored. In fact, the most frequent alternative raised by respondents was reducing carbon emissions:

> Why put large amounts of effort into reducing the amount of sunlight (which I think is a bad idea)? Put efforts into developing non-carbon based (this includes not using plant/food for fuels) sources of energy such as nuclear, solar (which would seem to be hurt by reducing sunlight), wind and hydro. Also why not put efforts into making our use of energy more efficient to reduce the amount used?
>
> Understanding we are in serious trouble with global warming, I support efforts to stall or retard advancement. However, rather than a band-aid we need to STOP at all costs the emissions.
>
> I think it's better to tackle a problem at the source (i.e. cutting down on emissions) instead of combating/counteracting the effects with more man-made interference.

These comments indicate that rather than proposing alternative technological solutions to SRM (such as other forms of geoengineering), many respondents insisted that the problem should be tackled by cutting emissions. Some respondents specifically regarded SRM as a "quick fix" or "technological fix":

> Whilst using this technology will we carry on researching in other ways on how to stop climate change for good? Or will we just look to solar radiation management as a quick fix solution?
>
> This seems like another quick fix to an environmental issue we have. We need to stop trying to solve problems with ways that could cause more problems. Remember DDT?

Several interesting ethical ideas arise here. One is the thought that SRM is an "easy option" allowing us to carry on living in the same ways while doing nothing about "fixing climate change for good." These comments call to mind Gardiner's (2006b, 2010; see also Hamilton, 2011a) discussions about the way climate change opens up possibilities for extensive moral corruption. A second idea is whether, as one respondent suggested, SRM is one of a series of cases where new technologies have been invented to "fix" a problem but have ended up "causing more problems" themselves.

This may point the way to further useful philosophical work. Although a broader philosophical literature already exists on the idea of technological fixes, this has not been applied in any detail to geoengineering. Scott (2011), in the context of agricultural technology, notes that despite the term "technological fix" being widely used as a derisive label, people "typically demonstrate an overriding preference for technological fixes" (p. 208). Indeed, a small number of survey responses displayed this exact preference:

> Why haven't you started yet?

> Please, please, please do geoengineering . . . we must have the technology to adjust our atmosphere to protect ourselves from any eventuality.

These types of responses raise questions about whether something being a technological fix *matters* ethically, particularly if the outcomes of using it are better than the alternatives. This, in turn, generates further questions about the ways geoengineering might be regarded from the perspective of different ethical theories—those where *outcomes* are what matter rather than those that focus on *principles* or *character*. More broadly, the idea of SRM as a technological fix provides an opportunity to explore contrasting attitudes toward technological innovation, both within and between different societies (Borgmann, 2006; Scott, 2011; see also Corner & Pidgeon, 2010).[3]

Whereas most respondents prioritized cutting GHG emissions, they disagreed about whether SRM should be pursued while working on emissions reductions or whether research on SRM should be abandoned altogether. Most geoengineering studies agree that SRM is no "solution," but here we see the same divide. Scientists such as Crutzen (2006) propose research on geoengineering while adamantly advocating for carbon mitigation; others, such as Kiehl (2006), reject the pursuit of geoengineering technology altogether, arguing that such high-risk alternatives should not be on the table at this point. Similar division also exists in the ethical literature. Gardiner (2010) comes close to rejecting SRM research, arguing that our attention should instead be focused on GHG mitigation. However, Preston (2011), as we have seen, argues that we might be best to regard SRM as a last resort in the case of runaway climate change. Given that this distinction seems to be found in public responses, the scientific literature, and the ethical literature, there may be opportunities here for climate experts and ethicists to engage the public in discussions about the basic permissibility of SRM and other forms of geoengineering and about risk-risk trade-offs between climate change and SRM, involving how such risks can be weighed and compared, and whether such trade-offs and comparisons are ethically permissible.

4. Distribution of Harms and Benefits

A number of respondents raised questions about the impacts of SRM on particular regions of the earth. Some were concerned about the impacts on themselves and on regions in which they have interests:

> What would be the expected changes in the weather—changes in the air, land, and water temperature—changes in the amount of sunlight, rain, snow, fog, occurring in the direct area of Canada in which I own a home?

But others raised questions about impacts on vulnerable or poorer regions of the world:

> What area of the world are you concerned with? Asia in particular has been hit with severe flooding but there seems to be more concern with coastal cities such as NYC so I question what part of the world will benefit.

> How severely will this technology change weather patterns in the poorer parts of the world?

These questions about regional impacts suggest that the public is concerned about possible distributive justice issues related to SRM. Modeling studies do indicate that sulfate particle injection could alter regional precipitation patterns, particularly in Africa and Asia (Ricke et al., 2010; Robock et al., 2008). This raises important regional justice issues, not least because those areas that have contributed least to creating the problem of climate change might be impacted most by this way of "fixing" it. These kinds of distributive justice concerns have been highly significant in the current ethical debate about climate change more generally but have not been much developed in work on geoengineering ethics (Jamieson, 1996; Gardiner, 2010; Royal Society, 2009). Yet this is clearly salient for members of the public. And the issues here are particularly important, as the distributions of harms resulting from SRM would be much more clearly *intentional* than harms resulting from climate change, raising different questions about moral responsibility and legal liability (Hamilton, 2011a).

Respondents were not only concerned about the distributive effects of SRM on present generations but on future generations as well:

> Will this kill my grand and great-grandchildren?

> My children and grandchildren are the future of this planet. We owe it to future generations to further explore ALL avenues of science AND restraint. We need to be much more aware and prudent, in our consumption and use of our natural resources.

These comments, anchored in the personal concerns people have for their children and grandchildren, indicates that respondents were also worried about the effects of SRM on the future. Whereas ethicists have focused on the long-range effects of climate change itself, not much yet exists concerning the potential impacts of SRM on future people. Although Gardiner (2010) discusses SRM and the future in some detail, his concern is primarily with the moral corruption current generations display by putting future generations in the position where they may have to act "evilly" by using SRM to protect themselves from dangerous climate change.

In expressing concerns about distributive and intergenerational justice as well as about the constraints on moral choices we might be leaving future people, the public has picked up on important justice considerations related to SRM. These concerns point to a pressing need for further studies of wide-ranging justice issues in the SRM context.[4]

5. The Role of Science, Trust, and Public Participation

Respondents were also concerned about how scientific research into SRM would be carried out and who would be in control of it. A number of questions were raised about the funding and conduct of research into SRM (including about the survey itself):

> What will be the benchmarks for success/failure, who will identify these criteria and how will these trials be funded?

> Who is paying for this research and who will benefit financially from the technology that emerges from it?

> Who pays you? How free are you to have an honest opinion? What environmental groups do you support? Are you asking because you value my opinion or because you need to spin the information to persuade me? What is a small scale research project and what effects will it have? Where will you do the research? Who is most at risk from this proposal? . . . how open are you to dialogue with independent media and environmental groups and the uninformed, non-academics like me? Will you call for round table discussions and offer all your data for perusal and invite all stakeholders to partake, including random people from every socio-economic sphere?

This last quotation also indicates a concern expressed by a number of respondents about the willingness of scientists to open their research up to public scrutiny and broader public dialog:

> Would the findings of the scientists be published in full for the public to study? Not to hide any information to the public like many wealthy corporations do in this country.

> I would like to know how long this research has been ongoing to reach this stage. Are we being told of all potential damage and dangers involved . . . Something should be done and I am in agreement with research but we should all be informed honestly throughout each stage. Who would benefit financially from this? Would companies vie for business? How would it be funded?

Along with calls for openness, honesty, and transparency, some respondents openly questioned why they should trust scientists on this issue at all:

Why should the public trust you?

Why do you think the public should trust your judgment?

How could you trust even yourselves to not use the science for personal gain and which government in the entire world would not use it for political gain?

These comments reveal widespread suspicion about scientific research into SRM operating at various levels: suspicion about motivation, suspicion about benefits, and suspicion about secrecy. Some of the comments suggest that these suspicions would be at least partially allayed if open and transparent scientific processes made research available to the public.

These types of concerns have certainly been anticipated in the major reports on geoengineering. The Royal Society (2009), Asilomar Scientific Organizing Committee (2010), British House of Commons (2010), Bipartisan Policy Center (2011), and Solar Radiation Management Governance Initiative (SRMGI) (2011) have all suggested principles that should govern research into SRM and ways in which public participation should be incorporated into the research process. The recent SRMGI report, published after this survey was carried out, explicitly raised issues about the funding of research and public perceptions of trust and liability. However, this is certainly a case where talk may be cheap. The announcement in the UK of a field trial of technology that could be used to deliver sulfate aerosols to the stratosphere in the future generated an international outcry in late 2011. Much of the outcry concerned the lack of public consultation about the research that led to the announcement (Macnaghten & Owen, 2011). This incident, which resulted in deferral of the field trial while further public consultation was carried out, indicates the importance both of keeping the public informed about scientific developments and perhaps even providing the public with the opportunity to help frame research questions, and goals (Stirling, 2005; Wilsdon & Willis, 2004).

6. Governance of SRM Technology

Alongside concerns about researching SRM, respondents expressed concerns about governance. Questions were raised about who would be in control of SRM technologies and who would be responsible if or when unexpected or negative outcomes occurred:

What industries would develop the technologies and who would be responsible for implementing them? What kind of national and international oversight would there be and with what authority?

Who would decide which part of the world would have their rain volume changed because of Solar Radiation Management and would that mean that some countries would need to become uninhabitable?

How will you predict where effects such as extra rainfall will occur? And if the extra rain exacerbates flooding somewhere, how will it be dealt with? Who will take responsibility?

These types of public concerns were widespread and raised questions about international relations and international responsibility. What governance regimes would deal with problems? And what would be an ethical way of doing so? Implicit here, as with the pursuit of scientific research, are questions about democratic participation in creating governance structures, in particular, how can those in hard-to-reach, but likely to be affected, locations be consulted? "With what authority" would there be national and international oversight?

Multiple parties, including the United Nations Convention on Biological Diversity (2010) and SRMGI (2011) and multiple publications, have discussed potential governance structures for SRM. However, these attempts to construct mechanisms for governing SRM do not carefully deal with all of the ethical issues that lie behind the worries expressed above. In particular, because of its intentional nature, SRM generates different questions of moral responsibility than climate change. If the implementation of SRM harms some nations while benefiting others, new international justice questions are raised. SRM will be carried out from a point source or sources; it will be a decision deliberately made by some group or groups; it cannot be attributed to people now dead, and so on. Ethicists should explore very closely the implications of these international justice questions raised by the governance of SRM.

The importance of public engagement is at least as pressing in developing the governance of SRM technology as research into the technology itself. There has already been some discussion of the ethical requirements here: Jamieson (1996) considers the need for democratic processes in the context of intentional climate change; Corner and Pidgeon (2010) describe upstream engagement as a potential mechanism for soliciting meaningful public participation in decision-making processes about SRM research and governance; and SRMGI (2011) discusses in some detail different forms of public participation in creating governance structures. But there is still work to do, and scrutiny is needed to ensure that espoused ethical principles of participation in decision making are actually implemented in practice.

7. Climate Denial

Finally, and in line with research on public perceptions of climate change, a handful of respondents questioned the underlying assumption behind SRM research, namely, that anthropogenic climate change exists at all:

> You are starting from a premise that is unproven, many would even say unfounded. That is, that human activity results in global warming (climate change). This idea is nothing more than a political maneuver to gain votes. Hard, peer-reviewed science indicates that the earth goes through cycles of warming and cooling. Nothing to be alarmed at.

> Why do you continue to waste money on charlatan theories and political grandstanding when overwhelming evidence suggests that the current climate changes are cyclical in nature and that carbon emissions have not been shown to impact climate nearly as much as the public has been led to believe?

It is important to acknowledge climate change deniers in this context for two reasons. First, as Jamieson (1996) notes, the prospect of geoengineering may create strange bedfellows. Those who oppose SRM (and perhaps other forms of geoengineering) as unethical may find themselves in an alliance with those who oppose SRM because they do not accept climate change at all. Alliances over particular moral questions for widely divergent underpinning reasons are not uncommon, with alliances between feminists and conservatives against pornography being an obvious example (West, 1987). But in this case, an alliance between climate change deniers and those opposed to SRM for other reasons may well be a fragile one. If SRM could be implemented without requiring any significant change to "business-as-usual," in contrast to what is demanded by significant GHG mitigation, politically and economically conservative climate deniers might easily shift to being SRM supporters. It is also clear that, right now, those who sincerely reject the science of climate change present another form of ethical opposition to SRM that has not been considered. This is an area where ethicists could inform future public engagement efforts aimed at exploring the relationship between climate denial and perceptions of geoengineering.

CONCLUSION

In this chapter, we have reproduced and discussed some of the key ethical concerns raised by the public about SRM and brought these concerns into dialog with ethical debates found in formal reports and academic papers. This dialog has led us to draw a number of conclusions.

First, and most generally, we have shown that ethical considerations are central to public concerns about SRM. This has emerged in other public engagements (Godfray, 2010; Parkhill & Pidgeon, 2011) but it is a point worth emphasizing. The core questions raised by geoengineering technologies in the public arena are, as the Royal Society (2009) anticipated, social, ethical, and political, rather than primarily scientific and technological. Over two-thirds of the responses to a very broad, open-ended question on an international survey focused on what we have classified as ethics issues. This indicates that ethical questions are at the forefront of people's minds when they think about geoengineering, and it is these questions that the public will want to see addressed by scientists and decision makers (Wynne, 2006). If SRM is to gain *public* acceptance, it will need to gain *ethical* acceptance. And at the moment, there are many public ethical concerns that would need to be allayed—if, indeed, they can be. This, of course, assumes that for SRM to be deployed it *should* be widely ethically acceptable. Some might contest this, but all the major reports produced so far (for instance BPC, 2011; Royal Society, 2009; SRMGI, 2011) emphasize that geoengineering technologies should be developed ethically.

Second, we have shown that, in many cases, the concerns raised by the public are closely related to those articulated in ethics literature. Unsurprisingly, public concerns about the ethics of SRM are often expressed more informally than scholarly ones, and though they emerge from personal life experiences and contexts, they are recognizably the same concerns. Because the academic ethics literature on geoengineering is relatively small, public concerns flag important areas for future work. For instance, a number of public comments raised questions about the distribution of the effects of SRM—who would benefit and who would be worse off—and what questions about distributive justice this raises. There has, so far, been little ethical work on these questions. Even if it is unclear at present what the effects of SRM might be, further research on the ethical significance of the intentional nature of SRM in contrast with anthropogenic climate change seems particularly important.

However, the traffic here is not all one way. While raising issues that ethicists should explore, public concerns should also be subjected to critical scrutiny. In cases where popular claims or worries harbor factual inaccuracies, reflect inconsistencies, or are otherwise problematic, ethicists should try to make alternative arguments available in the public sphere, perhaps by contributing to popular publications, public discussions, or participation in public engagements. Indeed, even where the majority of ethicists and the public appear to agree, there is still space for debate and objection. For instance, some members of the public viewed SRM as problematic because it intentionally interferes with natural processes. While this view is also held by the majority of environmental ethicists, it is, nonetheless, an argument that can be challenged on a number of grounds: for instance that "nature" has already been intentionally "interfered" with or that the whole human/nature distinction should be abandoned (Jamieson, 1996; Preston, 2011). Such arguments are not decisive, but bringing them into the public sphere could stimulate debate. Indeed, the salience of ethical concerns may indicate that the public would be interested in engaging in thoughtful and reflective treatment of these issues. Professional ethicists, then, should think about how to utilize intuitive ethical concerns raised by the public to facilitate and inform more complex and pertinent discussions about ethics and geoengineering in ways that are accessible and digestible.

There are, in addition, some arguments prevalent in the ethical literature about SRM that did not directly arise in comments from the public—for instance, the debate about whether geoengineering research itself creates a "moral hazard," causing people to be more reckless in terms of carbon emissions (Hale, 2010). Such arguments, in more popular form, deserve wider public debate and discussion.[5]

Our third and final conclusion concerns the considerable space the responses discussed above reveal for mutual learning at the intersection of social science and ethical reflection on geoengineering. Social scientists can help ethicists to continue assessing the concerns of publics, including public conceptions of ethical issues, and how such issues can be effectively discussed. Social scientists can also help to incorporate public engagements into ethical assessments of SRM technologies. Just as the public may raise new research questions via social science for ethicists, ethicists can inform future social science research in several ways, including considering what ethical questions particular geoengineering proposals raise (for instance, about distributive and intergenerational justice and about human relations to nature) and in assisting social scientists in facilitating public engagements that specifically target these questions. We conclude, then, that ethicists and social scientists can and should seek out ways to work together more closely on the social and ethical challenges posed by geoengineering for the benefit of public debate, scientific research, and policymaking.

NOTES

1. We have corrected minor spelling and grammatical errors and shortened some quotations.
2. See the chapter by Sandler in this volume [editor].
3. See the chapter by Scott in this volume [editor].
4. See the chapters by Hourdequin, Smith, Ott, Whyte, and Preston in this volume [editor].
5. See the chapter by Hale in this volume [editor].

The Cultural Milieu

Chapter Eleven

The Setting of the Scene

Technological Fixes and the Design of the Good Life

Albert Borgmann

Conventional wisdom assumes that on the stage of life the setting does not matter much; what really counts is the acting. That assumption is profoundly mistaken. Consider global warming. The fact of global warming and its likely consequences are so evident that it is hard to understand why people persist in their disastrous behavior. To understand and redirect their behavior, or at least to make their behavior less harmful, we need to understand the significance of the setting within which they behave and fail to act appropriately.

What makes a person act has been a concern of philosophers, social scientists, and politicians. The major question is whether people behave emotionally or act rationally, and the major shift in recent conventional wisdom has been from the belief that people are rational agents to the view that we commonly behave according to motives that are concealed from our rational awareness. We do not decide freely, we follow our desires (Brooks, 2009; Surowiecki, 2010).

A technological fix (as Alvin Weinberg, who developed the concept and coined the term, understands it) is a response to the failure of reasoning, the failure of rational arguments that are intended to change people's unacceptable behavior. Moved by their emotions, people behave inappropriately (Weinberg, 1967a). A technological fix changes the setting of their behavior in such a way that the behavior is no longer inappropriate. Solar radiation management (SRM) is a technological fix, based on the standard assumption that rational argument has failed to make people desist from their inappropriate behavior and on the distinctive assumption that the inappropriate behavior has intolerable consequences.

The technological fix is a species of a general cultural arrangement that assigns reason and emotion their appropriate functions in the design of the good life. The designer of the technological fix rationally arranges the circumstances of daily life so that people going with their inclinations are likely to do the right thing or so that the inappropriate thing is no longer such. The technological fixer implicitly recognizes the power of the setting in shaping or defusing people's behavior.

Setting aside for the moment the question of when a technological fix is justified, we are left with two general questions regarding the setting within which we play the roles of our lives: (1) Who in general should be the rational designer of the stage setting? (2) What kind of life should the setting support? Appropriate answers may not forestall SRM. The answers may fall on deaf ears and the responsibility for keeping the planet livable may have to be turned over to technical experts. But before we give up on placing responsibility where it properly belongs—with people and their civic responsibility—let us try to articulate a vision of the good life that has a claim on people's fears and hopes, and on their consent, so they will do the right thing in response to the challenges of global warming and global justice.

Global warming is here. The scientific evidence that it is occurring and that human activities are its main cause is overwhelming (Stern, 2010). Its injurious effects are already visible, and its catastrophic effects in the future are likely. But people in the United States seem to ignore all this. Just why they do and what can be done are questions that seem to have plausible answers (Moser & Dilling, 2004). Yet for all their plausibility, the answers lack a vital and hopeful center. So how to stave off a likely catastrophe? In 1967, Alvin Weinberg published an essay titled: "Can Technology Replace Social Engineering?" His eventual answer was "never." The essay has become a landmark in the philosophy of technology because it introduced the concept and the term of the technological fix. Intellectual landmarks are too often treated like mountains that are remembered only for their very peak. The entire substructure of valleys, ridges, trees, meadows, and trails sinks into oblivion. What is worse, at times it is a false peak that is remembered. Weinberg is sometimes wrongly remembered as a fulsome advocate of technological fixes. He does, however, give pride of place to social engineering—not a lovely term, for sure, but certainly referring to the right thing, to the work of social scientists, not necessarily excluding humanists and philosophers.

There are details where Weinberg was wrong or a little off target, but he did recognize an important and frequently neglected truth—that people react to the prompts of the material environment in typical and at times deplorable ways. Weinberg was well-aware of the influence of the social environment on people's behavior. He called the features of the social environment "social devices—usually legal, but also moral and educational and organizational"

(Weinberg, 1967a, p. 7). A technological fix, to the contrary, is a material device. Weinberg did not see, or at any rate did not discuss, the breadth and depth of these material devices that are always and already directing the normal course of people's lives. He was concerned with the insertion of particular novel technological devices, the technological fixes. Still he recognized at the very least that we direct people's moral conduct by shaping their material environment. If we change the latter appropriately, we improve or neutralize the former.

When he discussed the human response to the material or social environment, Weinberg again was insightful if imprecise. He equated the right behavior with "rational" behavior, "acceptable to society." Wrong behavior he saw as not rational and motivated by "immediate personal gain or pleasure" (Weinberg, 1967a, p. 7). Weinberg's view that we ought to act rationally, but do unfortunately behave emotionally, is ancient and widespread. It has come under attack recently, but we cannot judge the controversy until we have clarified some ambiguities.

How should we think of the relation between reason and the emotions? That they are not one and the same is evident from the rueful experiences we have had with our rational resolutions. We know what the right thing to do is. To use Weinberg's examples, be parsimonious in your consumption of water, judicious in your sexual practices, and peaceable in dealing with your adversaries. But our resolutions to do the right thing crumble under the weight of our feelings and desires.

Given the conflict of reason and passion, there are naturally advocates of reason and champions of passion. The former contend that reason should rule our passions by establishing moral norms we are to follow no matter our feelings. The opposing tradition takes emotions to be supreme and reason to be the servant or fool of the passions. The battle of the opposing views was evident in a recent issue of the *New York Times Book Review*. One book under review was Jonathan Israel's *Democratic Enlightenment* (McMahon, 2011). Its chief argument is that "applying reason tempered by experiment and experience, not anything based on blind authority, would bring vast social benefits" (Israel, 2011, p. 3).

The other book under review was by Robert Trivers and titled *The Folly of Fools: The Logic of Deceit and Self-Deception* (Horgan, 2011). As Trivers argues on evolutionary grounds, truth is not primarily found in reason and consciousness; rather "the information is preferentially excluded from consciousness and, if held at all, is held in varying degrees of unconsciousness." And why is this so? "The hypothesis of this book," says Trivers, "is that this entire counterintuitive arrangement exists for the benefit of manipulating others. We hide reality from our conscious minds the better to hide it from onlookers" (Trivers, 2011, p. 9).

But why should we believe Trivers? Is he telling us the truth or is he hiding it from us? Is his book a rational argument addressed to our consciousness or is it meant to fool us? Surely there is no way to answer these questions but by the canons of truth and reason, and I am sure Trivers wants to see his book so judged. The critics of reason are right in pointing out that there is no escape from the unconscious and from our emotions. But neither can we evade the deliverances of consciousness and reason. The relation of these two sides of the mind has to be the kind of equilibrium that emerges when reason accepts the lessons of the emotions and the emotions follow the instructions of reason. Without that balance, the instructions of reason will go unheeded, and we will remain shackled by popular emotions and left with the necessity of something like SRM. But it is important to realize that the equilibrium in question is not a moral fix solely for the purpose of preventing a technological fix. It is rather a basic moral requirement, and its importance comes clearly into view when we look at the attempt to entrust human well-being one-sidedly and entirely to emotions—the feelings of pleasure.

Hedonism, so-called, begins with a seemingly innocent and disarming admission: let us face it, pleasure is what we are after, however we may disguise our pursuit of it by protesting our means of pursuing it—politics, poetry, piety. So why not drop the pretensions and go straight for pleasure? Hedonists soon discover that the heedless pursuit of pleasure can produce a net balance of pain—much pleasure of wine tonight, more pain of nausea and embarrassment tomorrow. To secure pleasurable emotions, rational judgment is needed. Reason without emotion is arid. Emotion without reason is perilous. Reason gives us a wide and discriminating view of the world. It is superior to emotion in that we can reason about our feelings with precision and coherence. Emotions are superior to reason in that they engage us with reality deeply and decisively; but the ways we feel about reason are obtuse. Reason never acts without the momentum of emotions, without being fueled by ambition, curiosity, or enthusiasm. Emotions, unfortunately, can escape reason and cause us to do things that make reason weep.

And yet human flourishing requires more than a balance of reason and emotion. It requires us to consider the reality that we respond to consciously and unconsciously, rationally and emotionally. However articulate reason and however profound emotion may be, the world out there contains more features and greater depth than we could ever discover in the mind when we examine it by itself and draw up hierarchies of mental faculties and typologies of emotion. The constructive lesson we need to learn from all this is that the breadth of reason and the depth of emotion allow us to respond to the world in characteristically different ways. Reason gives us the broad picture of the world and the ability to shape it in the large. Emotions respond to the forces of immediacy and intimacy, to the pleasures and perils of the here and now. That is really Weinberg's lesson—reason tells us what kinds of settings have to be in place so that people feel motivated to do the right thing. But here too the balance of reason and emotion has to reflect what one has to teach the other. Rational arrogance is no less injurious to human flourishing than emotional indulgence. Let me use *intelligence* as a term to designate reason when it is attentive to the promptings of reality and the emotions and when it is devoted to human flourishing.

Who then is the rightful authority in the intelligent design of the setting in which we act? In a democracy, the final decisions are the people's. They are the ultimate authority in the shaping of the public sphere. But they cannot be its original author. Intelligent innovation does not bubble up from the population as a whole. It comes from inspired individuals and groups, and it is impossible to lay down norms for the quality and validity of innovation once and for all. Innovation must warrant itself. Once articulated, innovation needs advocates, people who can see the value of a new design and who are ready to assume the work of democracy—convincing a majority of voters that the innovative design deserves their approval. Intelligence qua creativity and ingenuity is needed to come up with a good design. It is needed in another way to help design become reality—intelligence also has to be the understanding of people's apprehensions and aspirations and the ability to appeal to those sentiments.

Two of the many issues that require changes in the setting of our lives are global warming and global justice. Changes have to address the two issues. Global warming requires structural changes of national and global infrastructures, of those large material devices that support and direct daily life. Global justice requires cultural changes so that the lives we lead are commendable—admirable and available to everyone around the globe. The structural changes, if carried out, will inevitably lead to cultural changes; that is the lesson of Weinberg's essay. Within the constraints that reduce the emission of greenhouse gases, we will by definition no longer act irresponsibly, but also lead a different kind of life. It is not intelligent, however, to leave culture at the mercy of structural changes. We have to see to it that life within the reformed structures is worth living. More important still, we cannot leave the structural changes at the mercy of current culture. We have done so for too long. Most of the required structural changes have been known and tested for a long time, but United States culture in the aggregate has been unmoved by them. It is this recalcitrance to infrastructural change that may necessitate the technological fix of SRM to mitigate the worst catastrophe.

Because the intelligent designs of a sustainable infrastructure have been laid down in general and tested in limited applications, all that is needed here are a few reminders. The Kyoto Protocol and its successors have been concerned with the global requirements of structural changes. The United States, a technologically advanced and environmentally profligate country, is a helpful setting for what implementation will have to look like (Dean, 2011).

There is, first, the need to make buildings and appliances more energy efficient. Second, the same is true of transportation. Cars have to be more fuel efficient, and their use has to be curbed and in part replaced by public transportation. Third, industry and commerce need to be more energy-efficient. Fourth, we have to capture more carbon dioxide through vegetation and perhaps by sequestering it underground. Fifth, the generation of electrical energy must become less carbon-intensive through better power plant design and more wind and solar generation. Sixth, agriculture needs to be reformed so less fossil fuel and water is needed (Bittman, 2011a; McKinsey & Conference Board, 2007; Webber, 2011).

This list largely follows the McKinsey and Conference Board Report of 2007. Its authors proceeded on the assumption that there would be "no material changes in consumer utility or lifestyle preferences." They do, however, concede that there could be "potential changes in utility that might result from energy price changes associated with pursuing the options outlined in our abatement curve" (McKinsey, 2007, p. 2). But in spite of the reassuring tone of the report, little of the vigorous action it had urged four years ago has been undertaken. The structural program has fallen victim to cultural incrtia.

How do structural and cultural changes affect each other? Evidently cultural changes are needed to break a path for structural changes. But what is culture? Culture is to a society what character is to a person. A human being today is a creature of many currents and functions. Especially in the United States, most of us are of mixed ancestry, have been inhabitants of many places, have been favored and burdened in diverse ways, and are playing a variety of roles. But in the end there is just that one person who is distinct from all others, exhibits a particular style of life, and is responsible for it. Everyone has a certain character. Is that true? Are we all persons of character? Evidently "character" has a descriptive and a normative sense. It is used to describe the inescapable distinctiveness each of us possesses. It is also used as a norm of the vigor and coherence of life that we all should strive for.

So with the culture of a society. The common life of the United States is composed of a great variety of persons, institutions, and activities. And yet it has a recognizable physiognomy, however hard it may be to describe. It is a visage we, as a people, are responsible for. And here too a normative aspect becomes visible. Our national countenance is or should be one of vigor and patience.

Instead, the cultural setting people typically find themselves in is a cocoon of troubled comfort. It is not an all-or-nothing container but more like an enveloping web that has become thicker over time. Smart phones are secluding us from our world and from one another. We seek comfort in the easily consumable information that filters into our envelope, and all too many of us, when troubled by the rumblings of crises and catastrophes, take comfort in readily available food. We dutifully show up at our jobs, but work is no longer a disclosure of the world. Labor is not exactly a cocoon of comfort, rather, it is much like a compartment where we do our part to maintain a system we do not understand. And most of our leisure time we spend in a capsule that is becoming more colorful inside and more opaque to the outside. It is drifting free of the rigors of reality and into the nowhere space of anything-anytime-anywhere. Whenever bracing changes in the debilitating comforts of food or information are proposed, the rhetoric of resentment and rejection easily beats back reforms. Structural changes as proposed in the McKinsey Report arouse fear of deprivation and discomfort, and we are not comforted by assurances that "no material changes" in our way of life are needed, "no change in thermostat settings or appliance use; no downsizing of vehicles, homes, or commercial space," to quote the McKinsey Report (McKinsey, 2007, p. 2).

If the fear of deprivation and discomfort is not met directly and hopefully, structural changes will either take the form of risky technological fixes or they will be forced on a resentful populace by dire necessity. The authors of the McKinsey Report understandably "have not attempted to calculate changes in utility that might result from energy price changes associated with pursuing the options outlined in our abatement curve" (McKinsey, 2007, p. 2). Precise figures and detailed scenarios would be hard to come by. However, some of the large structural features of change and their cultural consequences are traceable even now.

A national economy cannot do everything. It cannot both maximize the comforts of consumption and reform the infrastructure. If global warming is to be slowed, resources have to be diverted from the production of consumer goods to the reform of the underlying machinery of construction, manufacturing, transportation, agriculture, and energy generation. But reform of the infrastructure requires the deployment of political measures that are greeted with no less resentment than the prospect of changes in the material structures of life—the reform will not happen without higher taxes and more regulations, and whenever such measures are proposed, they are defeated by warnings that they will derail growth and end prosperity.

Poverty is conservative. The poor hang on to what is working. They do not have the margins of surplus that can cushion the risks and failures or experiments. American society is anything but poor, however. The U.S. Census tells us that average household income has risen from roughly $1,000 in 1901 to $50,000 in 2001. The expenditure share for "non-necessities" went from 21% in 1901 to 50% in 2002 and 2003 (U.S. Department of Labor, 2006, pp. 1, 57–59). Such figures give us a very rough outline of what people could do without. The "necessities" are food, clothing, and housing. Surely no one would say that everything else is unnecessary. Nor are all of the necessities truly necessary. A quarter to a third of all food is wasted. Forty-two percent of food expenditure is spent on expensive and mostly unnecessary "food away from home" (U.S. Department of Labor, 2006, p. 58; U.S. Department of Labor, 2009, pp. 1–3). Also, too much food is consumed. One thing is clear, however. Non-necessities and waste provide ample room for reform.

But what is likely to happen within that space? The involuntary curtailment of what is truly unnecessary was forced on American society by the recession of 2007 through 2009, and the necessity of doing with less gave a hint of what the cultural consequences of less waste may be. Americans said they were "increasingly becoming homebodies who are reading, knitting, cooking, watching television and playing board games." People took to crafts and went to the libraries (James, 2009). Not all of the effects of a recession are benign. Losing one's job is one of the most crushing experiences a person can face (Bok, 2010, pp. 20–21), and between that catastrophe and the wholesome engagements a recession fosters, there are many degrees of deprivation and despair. Still, there is a positive possibility in the beneficial changes of a recession. It needs to be brought into relief and put in context. In the best cases, austerity did not shred the cocoons of comfort but opened them to a more centered and grounded life. The home shifted from a terminal of consumption to a place of engagement. Family members recognized and engaged one another in activities that invite doing, talking, and sharing.

Consider cooking as a revealing exemplar. Contrast it with convenience food that has been carried or ordered in—a hamburger, fries, and some pop for everyone. The stuff appears essentially from nowhere but is engineered to occupy your desires. So all grab their portion, sink their teeth into it, and then begin to satiate that other hunger, the one for information; and the content of iPhones and iPads obliges that hunger with the alluring bits and pieces that appear from nowhere and commandeer your attention, just as the contents of the McDonald's box have been capturing your palate.

Cooking demands some awareness of the world you live in. You have to know and navigate it through the decisions you have to make—going to the farmers' market rather than the supermarket, selecting this lettuce rather than that, deciding to buy turnips rather than potatoes, and so on. Food like this is no more expensive than junk food (Bittman, 2011b), and it has the virtue of displacing the hidden machinery of McDonald's food engineering with the comprehension and competence of the cook. Home cooking, finally, replaces injurious junk food with a healthy meal (Szabo, 2004). Once in your home, your beloved, if not ready to help with the cooking, will at least be drawn into the kitchen by domestic noises and rising fragrances. If you have taken the trouble of shopping and cooking, you will not be shy to call your loved ones to the dinner table. That is actually the worst case. If your loved ones have cooperated in preparing dinner, they want to see it taken together.

And consider reading. Appearances to the contrary, it is a more engaging way of appropriating information than is watching (Ulin, 2009). The austerity of print forces our intelligence and imagination to supply what is needed to make a text come to life. Of course, whether such engagement enlarges our comprehension of the world depends on what we are reading. But a turn to libraries and books is another hopeful exemplar of reclaiming the width and depth of reality.

Home is the place where the blessings of competence are most readily imaginable and realizable. The home in turn opens up to public places of engagement, like farmers' markets and libraries. Vigorous domesticity needs the context of communal institutions such as concert halls, museums, libraries, playing fields, trails, and parks.

At this point I may be losing the intelligent advocates of structural and cultural change. My proposal may seem well-intentioned but too idyllic and controversial to deserve confident support. The good life, one may think, is a matter of subjective preference. We have no right to tell people, far less to constrain them in, how to live their lives. All we can do is warn them of impending catastrophes and inform them of necessary structural changes and the possible need for SRM. Venturing into cultural advocacy is arrogant and in any case unwarranted. So before I go on, let me reply to the skeptics and adversaries. The reply is in two parts. First, it is not possible for society to leave the question of the good life open and the decisions on what sort of life to live to the individual. Sure there is room for individual choice in our society; let it be noted gratefully. But for most people the constraints on choice are narrow and strong. Forty years ago that claim may have required subtle analysis and complex arguments. But today the incredible power of at least some of the material and social constraints is too obvious to ignore. Have two-thirds of the population decided individually to become overweight or obese or have they responded to the material changes in the setting of their lives that were set in motion over the past decades? Are people now deciding to distract themselves with junk information, or are they responding to the technological innovations that have been promoted by government and business? A host of sorrowful reports testifies to the tyranny of electronic devices (Barker, 2009; Carr, 2011; Harris, 2011; Meece, 2011), and there are also testimonies to the pleasures of throwing off the yoke of the Internet (Cohen, 2012; Iyer, 2012; Sotos, 2009).

The second part of my reply is this. If normative or traditional arguments fail to convince you that a life of social engagement and physical skill is the good life, consider the valid and reliable findings of the social sciences. Inevitably they vary somewhat from time to time and research to research, but the converging evidence of much research is that interacting directly with people and skillfully with tangible things are most conducive to what is variously called well-being, human flourishing, or enduring happiness (Bok,

2010, p. 28; Juster & Stafford, 1985, p. 336; Putnam, 2000, pp. 263–65, 334, 402–14; Seligman, 2002, pp. 1–161). Just as important, social science evidence shows that the setting of our society systematically diverts most people from the good life (Easterbrook, 2003; Lane, 2000; Myers, 2000).

The economy then, needs to be directed by degrees toward a life of competence and comprehension. There have to be more bicycles and buses and fewer cars, more books and fewer screens, more skis and fewer snow machines, more recorders and fewer iPods, more canoes and fewer jet skis, more farmers' markets and fewer supermarkets, more handcrafted furniture and less mass-produced clutter. These changes cannot be left to individual decisions. People respond to the setting of their actions. "Reason tempered by experiment and experience," to use Jonathan Israel's expression, tells us that changes in the material setting require the support of Weinberg's "social devices." People's behavior responds to changes in taxes and prices. "Every 10 percent increase in cigarette prices reduces youth smoking by about seven percent and total cigarette consumption by about four percent," to cite one instance (Campaign for Tobacco-Free Kids, 2012). To cite another, "after the 1979 spike in crude oil prices, US gasoline consumption dropped for four years, but then rose again when fuel prices plummeted in the mid-to late-1980s" (Gold & Campoy, 2009).

There are many appropriate social devices to create these changes. Activities that are injurious to human flourishing and the environment have to be taxed heavily—junk food, driving, energy-intensive agriculture, fossil-fuel electricity generation, and scattered housing among them. Enterprises that favor competence and comprehension have to be promoted—physical activity and music in K-12 education, rigorous general education in higher education, local labor-intensive industries, and walkable towns with inviting opportunities for physical and cultural activities among those. A crucial challenge is to distribute the burdens of change fairly. Domestically, for example, if the United States imposes or raises taxes on fossil fuels, the rich need to pay higher income taxes so that the fuel expenses of the poor can be subsidized. Internationally, we have to realize that, however green our infrastructure may become, it will rest on industrial processes of the past that in large part have brought us to the present crisis. Hence, fairness requires that we share the benefits of our past harms to the atmosphere with the countries that are now in the process of building up their infrastructure.

There are reasonable arguments that the setting of our lives can be so restructured that doing what is good for human flourishing and right for global justice and a stable climate becomes the natural thing to do. It could take a very long time until our lifestyles are simplified down to the level where the standard of living of the poor around the globe can be brought close to ours, and it will take just as long to bring the global culture in line with what the planet can sustain. But a structural and cultural program that will make SRM unnecessary is possible.

Yet contrary to Hegel, the rational is not real. Intelligence requires more than appeals to reason. We have to show that it feels good, and is good, to do the right thing. The good life, the one that makes people deeply and enduringly happy, is one of skillful engagement with the world and with one another. There are certainly ways in which the good life could be costly to the environment and to the poor countries. Imagine an elite of the good life, whose every member has a string of horses to play polo and a tracker pipe organ to play authentic Bach. The life of that elite would be skillful, grounded, and centered. But it is not the kind of life that could be shared across the globe. Our lives will be no less good and fulfilling if instead we take up running and the guitar. This kind of life would be more profoundly good because it would be free of the reproach of selfishness and regardlessness. Barring a cataclysm, the future will either see something like SRM or it will bring the good society. But there will not be a good society without intelligent and confident advocacy of the good life.

Chapter Twelve

Between Babel and Pelagius

Religion, Theology, and Geoengineering

Forrest Clingerman

Climate change, and the need for a comprehensive and drastic response to it, is a well-publicized issue. Yet because it is so pressing and all-pervasive, climate change is not simply a scientific issue—it is also an issue of our deeper sense of human values. Proposed responses to climate change emerge out of the values held by individuals and societies. The tensions in our approaches to climate change arise from a conflict of interpretations of how to act in the world and of whom we are meant to be—that is, our approaches to climate change arise from implicitly *religious* paradigms.

Deeper than the explanatory power of scientific models, the groundwork for our understanding of climate change rests on spiritual and ethical impulses. Christian theologian Mark Wallace comments, "science cannot assert, as religion can, that the Earth is *sacred* and therefore deserving of our protection" (2011, p. 12). Wallace is making two interrelated claims, both of which should be taken seriously. On the one hand, religious language conveys a unique meaning that is missing in exclusively scientific explanations. On the other hand, religious beliefs frame how individuals and communities approach the debate. These two claims are dynamic and mutually reinforcing. Because religion adds a dimension of meaning to environmental debates—a dimension predicated on the idea that the "sacred" consummates or transcends the mundane—religious communities create new frameworks for understanding the world. At the same time, religious frames of reference attune religious communities to different levels of possible meaning, which are expressed in religious language.

Certainly the influence of religion has broad implications for a pluralistic society, particularly when religion surreptitiously colors our perception of things not commonly taken to be "religious matters," such as climate change. Mike Hulme convincingly explains that religion impacts the societal discussion of climate change precisely because "[o]ne of the reasons we disagree about climate change is because we believe different things about our duty to others, to Nature and to our deities" (2009, p. 144). Religious beliefs are at the core of individual and social self-identity, even when we are unaware of this influence. Religion, therefore, influences how science and policy are understood in the broader public square. It defines ethical responsibility and the ontological structure of the cosmos for its adherents. Certainly policymakers and scientists should be wary of replacing scientific study of climate change with religious sentiment, but they also must remain aware how much debate occurs within and between religious communities about the *practical* implications of doctrine. Asking what resources religious reflection offers to the discussion of the climate, then, poses an opportunity for understanding the ethics of geoengineering. How does our religious interpretation of who we are—and what the nature of the world is—influence our perspectives on geoengineering?

This essay suggests how religion might impact the debate concerning the ethics of geoengineering, specifically the more technologically complex strategies of solar radiation management (SRM). SRM, as one form of geoengineering, is oriented toward large-scale climate modification projects that are complex, technologically-dependent, and quick-acting manipulations of the earth's albedo. Morgan and Ricke explain,

> If the fraction of sunlight reflected by the earth back into space (the albedo) is slightly increased, then the amount of sunlight that is absorbed by the earth system is slightly reduced and the planet is cooled. . . . Large explosive volcanic eruptions clearly demonstrate this when they add large amounts of fine reflective particles to the stratosphere. Humans could do similar things to increase albedo. . . . (2010, p. 5)

SRM is intentional, anthropogenic climate change that attempts to affect global or regional systems in order to increase slightly the reflectivity of the earth's atmosphere.

This definition of SRM highlights a tension between the artificial and the natural. Certainly some SRM proposals mimic natural processes; Morgan and Ricke point to particles from volcanic eruptions, for example, as something that could be replicated. The natural, as it is seen here, "refers to whatever exists which is not the result of deliberate human intervention, design, and creation. . . . The natural comes into existence, continues to exist, and goes out of existence entirely independent of human volition. . . ." (Lee, cited in Preston, 2011, p. 461). Thus even if SRM proposals copy natural processes, they are not themselves natural. Rather, SRM projects assume some degree of control over natural processes. SRM projects are artificial, especially if the artificial is something that "embodies a human intentional structure" (Preston, 2011, p. 461). SRM projects presume high levels of scientific acumen, technical aptitude, and human decision making. That is to say, they apply a form of knowledge about natural processes for human ends and purposes, which results in an artificial control of the climate. SRM, indeed, is the opposite of a natural process—without human direction and ongoing input, the intended amelioration of climate change through SRM will be temporary.

Clearly tied to human intention, SRM emerges from whatever frameworks that we (individually and collectively) use to assess human responsibility and interpret human meaning. For example, the confidence to undertake such projects implies certain qualities of human capability, purpose, and value. Similarly, dismissals of the viability of SRM projects assume certain traits about human individuals and society. This paper suggests that some of these assumptions are best investigated within the realm of theological reflection. For starters, any opinion of SRM offers an interpretation of our values and what it means to be human in the world. Theological reflection has much to contribute to our interpretational approach to these subjects. Furthermore, as Keith et al. write, "SRM research should not be entrusted exclusively to either its proponents or its adversaries. Instead, there may be value in a 'blue team/red team' method, in which one team is charged to propose an approach that is as effective and low-risk as possible, and the other works to identify all the ways in which it can fail" (2010, p. 427). With so much at stake, then, the discussion of SRM must be as vigorous and wide-ranging as possible—and this includes reflecting on how religion provides a lens on the meaning of who and what we are.

How might we explore the relationship of SRM, religion, and the ethics of geoengineering further? In this essay, I approach this task by first defining the task of theology and explaining its relationship to the discussion of climate change. In the second and third sections, I will suggest two opposing poles of theological narratives for interpreting SRM, using the metaphors of the Tower of Babel and the early theological position of Pelagianism. Fourthly, I conclude with a few comments on how this range of theological responses might influence the discussion of the ethics of SRM. Each of these sections provides a brief sketch, hinting at avenues for further discussion on our interpretation of geoengineering. Broadly, we see that on the one hand, religious communities can draw on their own theological resources to assess SRM; on the other hand, we can also see how proposals focused on the deliberate manipulation of the climate suggest a theological or spiritual perspective. Religion, in sum, has a place in how we discuss the ethics of geoengineering.

RELIGION, THEOLOGY, AND THE DISCOURSE ON CLIMATE

The connection between religion and the "environmental crisis" has been debated for decades. For many scholars, this debate began in earnest with Lynn White's famous essay, "The Historical Roots of Our Ecologic Crisis" (White, 1967; see also Whitney, 1993), in which White argues that medieval Christianity formed the intellectual framework (and must be given a significant amount of blame) for the modern treatment of the natural world. In response, others have argued for a more complex interrelationship between religious belief and our attitudes toward the environment. Much of the discussion focuses on the value of historical traditions—usually Christianity—in confronting contemporary ecological problems. For example, H. Paul Santmire (1985) characterizes the history of Christian theology as having an "ambiguous promise" toward nature, and other theologians (e.g., Edwards, 2006; Schaefer, 2009) chronicle how the history of theology is a strong resource for environmental responsibility.[1] Regardless of one's position on *how* religion has influenced our understanding of nature, there seems a general agreement *that* religion influences our treatment of our environment. Religion and spirituality are undeniably part of the fabric of human life; for human individuals and for entire societies, religious practice is a lens through which we understand ourselves and the world around us. And this lens is pervasive: for instance, Gallup reports in a poll from the summer of 2011 that 92 percent of U.S. adults believe in God (Newport, 2011). With that in mind, the present section defines the task of theology—the systematic reflection on religion and spirituality—and suggests that the theological dimensions of SRM must be attended to in the debate.

The connection between religion and climate is neither modern nor limited to a single culture. Rather, the connection between human actions, divine will, and the changing climate is at the heart of religion itself. As Simon Donner notes, "[t]he weather . . . is up to God." He continues, "[f]rom ancient mythology to indigenous belief systems to modern organized religions, weather phenomena are believed to be explicitly controlled by the sky-dwelling gods. The weather is central to creation myths. . . . It is also used to discipline or reward people for prayer" (Donner, 2007, pp. 231, 233). Especially in agricultural and pre-modern communities, weather significantly impacts human culture and well-being; the sheer unpredictability of climate often necessitates a religious response. As a result, *religion can be counted as undertaking the first attempts at geoengineering!* It has used "technologies" such as rain dances; ritual sacrifice and offerings to induce favorable weather for safety or agricultural needs; communal prayer and acts of repentance in response to adverse weather; and rituals related to harvest. Such practices continue: one high-profile example is when the governor of Texas,

Rick Perry proclaimed April 22 to 24, 2011, to be days of prayer for rain as a response to drought and wildfires. Furthermore, climate influences religion. Religion is an outcome of one's climate, especially insofar as one's climate presents a context for religion and religion offers an opportunity to exert some control over its power. As a result, theologian Sigurd Bergmann writes, "climate change challenges and changes images of God and the sacred and their corresponding sociocultural practices" (2009, p. 98). For Bergmann, "climate change" defines the material landscape through which our spirit and values are manifested.[2] If that is the case, our ability to manipulate the climate has deep religious implications. In a myriad of ways, then, SRM is tied to religion.

To say that geoengineering is a religious issue is different from saying it is a moral issue, though undoubtedly these are related. This becomes clear with an example. In discussing the moral hazards associated with geoengineering, David Keith writes, "The root problem is simply: Would mere knowledge of a geoengineering method that was demonstrably low in cost and risk weaken the political will to mitigate anthropogenic climate forcing?" (2000, p. 276). He goes on to mention that ethical arguments tend to be pragmatic: the most common ones include claims on the slippery slope of using geoengineering, the fact that geoengineering is a technical fix rather than a solution to the problem, and that geoengineering is unpredictable. While such arguments might be important, Keith concludes, "These concerns are undoubtedly substantive, yet they do not exhaust the underlying feeling of abhorrence that many people feel for geoengineering" (p. 277). Moral and ethical argumentation offers reflection on one important way to frame the discussion but it does not reflect upon—or critique—the overarching worldviews or frameworks through which such questions arise. What might be the cause of that "underlying feeling of abhorrence"? Religion is what remains below ethical analysis—and simultaneously religion is what animates it. Moreover, religion can be a compass for other facets of human life, such as our sense of value, beauty, and the structure our lives take.

More strongly, then, we can state that our understanding of geoengineering is impacted by religion, insofar as religion provides an overarching framework for identifying and understanding what is meaningful. Religion acts an interpretive key for humans, providing a structure through which to understand ourselves, the world around us, and what we deem sacred. Certainly religion includes institutions, rituals, established authorities, and doctrines. But more fundamentally it begins with a core sensibility of the meaning of the human condition together with the relationship between humanity, the world, and God. Thus Friedrich Schleiermacher, the so-called "father of modern theology," discusses religion in terms of a "feeling of absolute dependence" (1989, p. 12ff.) and a "sensibility and taste for the infinite" (1996, p. 23). This feeling defines our place as finite creatures over and against the infinite, according to Schleiermacher. Such a feeling does not occur in the abstract but in our interaction with the universe in all its complexity and majesty.

Likewise, Paul Tillich defines religion as dealing with one's "ultimate concern." Tillich explains:

> The religious concern is ultimate; it excludes all other concerns from ultimate significance; it makes them preliminary. The ultimate concern is unconditional, independent of any conditions of character, desire, or circumstance. The unconditional concern is total: no part of us or of our world is excluded from it; there is no "place" to flee from it. The total concern is infinite: no moment of relaxation or rest is possible in the face of a religious concern which is ultimate, unconditional, total, and infinite. (1951, pp. 11–12)

For Tillich, this definition is broader than institutional forms of religion; nationalism, certain economic structures, and ideologies might function as one's "ultimate concern," or what concerns them ultimately. Since *everyone* has something that is of ultimate concern, everyone has a religious impulse. We all have the capacity for a "feeling of absolute dependence." *Religion*, in sum, *offers a ground on which one's sense of the world stands and a framework through which one interprets meaning.*

Theology, in turn, is the systematic reflection on spirituality and religion. Simply put, theology is "an ordered body of knowledge about God" (Hill, 1987, p. 1011) But this basic definition does not illuminate the extent to which theology is concerned with identifying and explicating how we interpret the world and ourselves in light of what concerns us ultimately. Moving beyond a pre-reflective "feeling of absolute dependence," theological thinking attempts to outline beliefs, doctrines, and philosophies that result from our religious experience. In other words, theology has a distinctly critical and interpretive task: it makes explicit the meaning of the ontological structures of finite and infinite as well as of the relationship between self and other. Using Tillich's definition of religion, this suggests that theology seeks to determine the "depth of being": what is ultimate and what is preliminary, and why. This means interpreting our "Godtalk"—a way of translating "theology" as the task of putting the knowledge of God into words and concepts—in the concrete contexts of human existence. In almost every case, theologians conceptualize the narrative we tell about ourselves by reflecting on the human encounter with what Rudolph Otto called "the Holy" (1958) and Martin Buber considered the "Eternal Thou" (1970).

The theological task can be applied to environmental questions, insofar as religion is a lens to determine value and meaning of natural and built environments. Theologians systematically reflect on how religion provides a framework to understand what might be "natural," particularly if the natural is connected with moral goodness, divine intention, and the role of humanity.

Theology also includes an investigation into the meaning and purpose of human existence within the natural world, especially in relation to our use of and inhabitation within a divinely-made "creation."

We can now define the place that religion holds in the discussion of SRM: religion adds narrativity and meaning to the discussion of climate change. Religious faith cannot provide scientific explanations or models through which to understand climate change. Neither is it able to offer technical fixes or engineering plans for SRM. What religious traditions do offer are narrative frameworks through which to integrate diverse, otherwise eclectic visions of the world. Jakob Wolf and Mickey Gjerris point to one illustration of the connection of religion and narrative: a recent UN Development Programme and the Alliance of Religions and Conservation "White Paper on the Ethical Dimensions of Climate Change." The White Paper assumes that religions are necessary conversation partners for not only practical reasons but also for reasons based on philosophical and ethical positions. Wolf and Gjerris cite that the report notes:

> [T]he emphasis on consumption, economics and policy usually fails to engage people at any deep level because it does not address the narrative, the mythological, the metaphorical or the existence of memories of past disasters and the ways out. The faiths are the holders of these areas and without them, policies will have very few real roots. People need to understand why certain archetypes, myths and stories work and others do not. (2009, p. 122)

Wolf and Gjerris summarize by saying " . . . faiths are—by virtue of their stories, myths, metaphors and images—able to engage people at a far deeper level than politics and economics" (2009, p. 122). For many, religions are not an ad hoc collection of doctrines and rituals unrelated to science. Instead, religion provides the predominant framework—a story—through which science, policy, economics, etc., are understood.

Because religion can act as a narrative compass, theology also has a distinct place in the SRM debate. As a systematic reflection on religious experience—and on the stories and meanings that emerge from established religious traditions—the discipline of theology offers a means to analyze the religious understandings that infiltrate the discussions of SRM, especially when these discussions appear in the public square. Theological reflection interprets religious experience, thereby providing an analysis that translates confessional discussions into a dialogue partner for public intellectual discourse. Parallel to this, theology translates aspects of this dialogue for individuals and religious communities, thereby offering a reasoned discourse for people of faith to understand the implications of SRM on their view of the world. By analyzing the religious questions of the human condition, the nature of the world, and the meaning of the sacred, theological reflection has much to offer to the SRM debate.

A NEW TOWER OF BABEL

How does our religious interpretation of who we are and what the nature of the world is influence our perspectives on the ethics of geoengineering? In fact, there is no single "religious response" to SRM. Different religious traditions have different authorities, rituals, scriptures, historical contexts, and theological commitments. These result in a dialogue between different approaches to technology, politics, and environmental concerns, for example. Individuals who claim no allegiance to a tradition complicate this dialogue even further.

Rather than attempt to catalog how several religious traditions might broach the subject of geoengineering and the effects these traditions have on the global debate, I suggest a continuum of responses that might emerge in the coming years. This continuum will be developed in conversation with one tradition: Christianity. While there are drawbacks to limiting the discussion to one religious tradition, there are two benefits that justify this focus. First, Christianity presents an optimal case study to explore religious narratives. This case study highlights the religion that has significantly influenced western society, including Western politics, ethics, and science. Second, the lessons we can draw from Christianity not only speak to western society, but these lessons also have elements that transcend Christianity and the West. Other traditions are the basis for similar discussions that can both learn from and raise questions concerning the admittedly preliminary discussion presented here.

In this section and the next, I offer two poles of the continuum of Christian responses to SRM: the Tower of Babel and Pelagianism. These two poles are the outcome of different theological models that interpret the human condition. If theology offers a framework for interpretation of the world, this framework results in certain challenges to geoengineering proposals: what is our place as humans—to manage creation, or to let it be? What might it mean to "have dominion" or be "stewards" (both phrases resonant to Christian audiences and diverse theological circles)? How does the theological recognition of human fallibility influence our interpretation of geoengineering?

In offering these possibilities, it is important to recognize that Christian communities and theologians have (as yet) said little about geoengineering. In the few academic works by theologians and Christian ethicists about climate change, none have given significant space for a discussion of geoengineering.[3] Likewise, no major Christian denomination has offered a position paper or theological rationale that responds exclusively to the questions of SRM or geoengineering, though there are numerous denominational statements concerning the human response to climate change. The following, therefore, suggests ways that the religious and theological discussion might be framed when Christian communities begin to approach the topic of SRM more seriously.

One theological model that can be applied to SRM is a reenactment of the Tower of Babel. A story from Hebrew and Christian scriptures (Gen. 11:1–9), the Tower of Babel has a place in western culture through written texts, works of art (most famously Pieter Bruegel the Elder's painting), and popular imagination. Briefly, the story concerns a supposed time when all humans shared one culture and one language. After settling the land of Shinar, humans said to themselves, "Come, let us build ourselves a city, and a tower with its top in the heavens, and let us make a name for ourselves, lest we be scattered abroad upon the face of the whole earth" (RSV, Gen. 11:4). In response, God "confused" them by scattering humans and offering a multitude of languages. The tower and its city remained unfinished.

There are interesting parallels between SRM and the Tower of Babel. For instance, both occur on behalf of the entire human population, led by cultural elites. Both show attempts by an advanced civilization to harness the pinnacles of technology for a hitherto unknown mastery of their surroundings. Both are undertaken to safeguard human civilization: the rationale of the Tower is to avert being "scattered abroad upon the face of the earth," while SRM is meant to avert the worst effects of climate change.

Most significantly, both can be interpreted through a narrative centered on the hubris of humanity. This narrative starts with the acknowledgment of human sin as a refusal to accept one's human limits. Reinhold Niebuhr influentially explained this claim:

> Man is insecure and involved in natural contingency; he seeks to overcome his insecurity by a will-to-power which overreaches the limits of human creatureliness. Man is ignorant and involved in the limitations of a finite mind; but he pretends that he is not limited. He assumes he can gradually transcend finite limitations until his mind becomes identical with universal mind. All of his intellectual and cultural pursuits, therefore, become infected with the sin of pride. Man's pride and will-to-power disturb the harmony of creation. . . . The religious dimension of sin is man's rebellion against God, his effort to usurp the place of God. (1964, pp. 178–79)

Both the story of Babel and SRM can be interpreted as paradigmatic examples of this rebellion. The Tower of Babel shows the sin of humanity in attempting to "make a name" for itself at the expense of God's authority. Likewise, interpreting geoengineering as an act of hubris rests on a narrative of how SRM proposals attempt large-scale, *artificial* intervention. To assume that human intentionality—one of the defining qualities of the artificial—can and should exert such an unprecedented power over climate runs counter to a religious sensibility that separates the domain of human control and the majesty of God. Geoengineering is "unnatural," not simply because it artificially controls the climate, but because it means that humans attempt to wrest control of creation from God. This narrative of SRM shows how humans meddle with what properly belongs to God. This will lead only to division, destruction, and human failure.

In this narrative, what is natural (and therefore religiously "appropriate")? The "natural" is what is ordained by God, who created a meaningfully ordered cosmos. To once again quote Donner, "The weather... is up to God." This control extends further than the sky. In Michael Northcott's discussion of energy use and climate change, he implies that Christian ethical response to climate change is to live "naturally" and on the human scale (something that is analogous to what Preston has called the "presumptive argument against geoengineering in environmental ethics" [2011, p. 461]). Northcott writes, "The reductionist account misses this complexly holistic character of the relational cosmos, and so removes the sense that what science calls nature has authority, and expresses constraints on what humans may do to it" (2007, p. 196). In response to this claim, SRM can be interpreted through sin: if the human place is defined as living within the order of God's creation, SRM can be interpreted as an audacious manipulation of the very order and foundation of the cosmos itself. Indeed, within many world religious traditions the divine is associated with the most basic "stuff" of the world, especially the sky or the heavens. Seen in this light, SRM is not an attempt to be "stewards of creation"—the theological role given to humanity in the opening pages of Genesis—but rather is a provocative tampering that is an unnatural distortion of the idea of stewardship. After all, SRM is an attempt to control what has been mythically understood as the realm of God. "In literature and myth, only gods and magicians had access to controls over the elements. But in the twentieth century, serious proposals for deliberate modification of weather and/or climate have come from engineers, futurists, or those concerned with counteracting inadvertent climate modification from human activities" (Schneider, 2008, pp. 3843–4). From a certain theological point of view, therefore, SRM becomes a contemporary Tower of Babel—humans are once again attempting to "make a name for themselves" and avoid a foreseen calamity.

With this in mind, there is a great possibility that many Christian communities will express a great wariness toward geoengineering. Expressions of disapproval already can be found in the literature around SRM and geoengineering, hinting at the need for a theological excavation of the discussion surrounding SRM. In an essay that spells out some of the difficulties surrounding geoengineering proposals, Alan Robock has succinctly listed many of the concerns and difficulties with geoengineering, two of which have direct bearing on the present discussion. The first that he acknowledges is the inevitable results of human error:

> Complex mechanical systems never work perfectly. Humans can make mistakes in the design, manufacturing, and operation of such systems. (Think of Chernobyl, the Exxon Valdez, airplane crashes, and friendly fire on the battlefield.) Should we stake the future of the Earth on a much more complicated arrangement than these, built by the lowest bidder? (2008, p. 17)

Robock brings up another problem: the complexity of asking who is responsible for controlling and maintaining the global thermostat. He rightly wonders who has the moral authority to judge these situations, which suggests a broader, more theological claim about the human condition: we are fallible creatures, prone to corruption and error. Moral authority, in this theological narrative, does not rest on individuals or even particular human societies. Rather, many Christian (and other theistic) communities would place the final moral authority in the hands of God. These communities would see SRM—given the global scale, the technological subjection of the natural, and the pitting of the needs of humanity against the divine plan—as an attempt to circumvent this claim.

When SRM is read theologically as a reenactment of the Tower of Babel, the appropriate Christian response is to live within the limits of the human position as "creatures" within creation. Humans are human, God is God, and the Tower of Babel is a reminder that we must remember the difference. For some, this might indicate that it is impossible for humans to be at all responsible for climate change (this leads to some conservative Christian arguments against the claim that it would even be possible for humans to change the climate). For others, however, the story of Babel points toward a more nuanced understanding of the human condition and the possibility of knowledge. Humans are finite and prone to error but nonetheless have responsibility for the earth as prescribed by their relationship with the sacred. SRM risks divine condemnation and wrath.

A theological position that attempts to balance humility and responsibility can be reinforced by certain reactions within the scientific community. For example, Stephen Schneider admonishes, "Rather than pin our hopes on the gamble that geoengineering will prove to be inexpensive, benign and administratively sustainable over the centuries—none of which can remotely be assured now—in my value system, I . . . would prefer to start to lower the human impact on the Earth through more conventional means" (2008, p. 3858). That is to say, the appropriate response is to know our place, given the priorities of "our value system." Furthermore, "Claims that the imperative of development cannot be impeded by the prospect of global warming are greeted with the assertion that creating inadvertent damage to nature is bad enough, but deliberately attempting to manipulate the climate just to let our old habits prevail is a violation of stewardship and an ethical transgression against the natural world" (Schneider, 2001, p. 417). It is not science that speaks of "stewardship" and "transgressions" against creation—it is a Christian worldview.

By noting the metaphor of the Tower of Babel, we see how geoengineering can bring out certain Christian theological claims, especially when we interpret SRM research as abrogating the limitations defined by value and meaning systems. If SRM is a rebuilding of the Tower of Babel, it means SRM goes against the story built around a particular ultimate concern of human existence in the world. The result, this religious story tells us, is dire: "Before we start geoengineering we have to raise the following question: are we sufficiently talented to take on what might become the onerous permanent task of keeping the Earth in homeostasis?. . . We could find ourselves enslaved in a Kafka-like world from which there is no escape" (Lovelock, 2008, p. 3888). When overstepping our bounds punishment is inevitable, we might ominously conclude.

THE OPTIMISM OF PELAGIANISM

The focus on human fallibility and the humbled place of the species is not the only Christian narrative that can be applied to SRM. In contrast, an alternative theological model springs from a guarded optimism of humanity. If the previous model takes the biblical imagery of the human fall into sin and the pride of the Tower of Babel, the present model suggests different biblical imagery: humans are created in the "image and likeness of God" in Genesis 1. But equally, the present model rests on the ideological framework of scientific progress. Like traditional religious denominations, the proponents of geoengineering offer a way of interpreting our place in the world, and implicitly this perspective is also a basic outline for "what concerns us ultimately." The model that serves as the second pole concludes that geoengineering proposals are rooted in a rudimentary crypto-theology and are centered on the human capability to know, a claim of responsibility, and the potential goodness of the human spirit. Interestingly, like the story of the Tower of Babel, the present story focuses on the centrality of the human being in the midst of creation. But in this case, human capability—and not human fallibility—is the foundation for the narrative. This position makes strong claims regarding the human ability to overcome the problems of the world. As Stephen H. Schneider writes:

> . . . many have claimed that the anticipated several-fold increase in greenhouse gases—and associated sea-level rises, intensified hurricanes, or drought and flood stresses—*can be largely overcome by human ingenuity* [emphasis add-

ed]. Their optimistic vision depends greatly on what has been called "geoengineering" and has more recently been relabeled Earth systems engineering. (2001, p. 417)

If the Christian story is one of humans in relation to the world and the sacred, then this story explains how humans deserve a dominant position in the cosmos and have the will and ability to control it wisely.

To connect this view with a metaphor, I suggest the theological position of Pelagianism. Pelagius was a theological opponent of Jerome and Augustine in the fifth century; while Pelagius' view was considered heretical, the purpose of making a comparison between SRM and Pelagianism is not to suggest SRM is "heresy." Instead, we see how SRM contributes to a possible theological narrative of the human relationship with the world. The central controversy between Pelagius, Augustine, and Jerome concerned the extent to which humans are infected by "original sin"—the sin and corruption of Adam and Eve's disobedience against God. Original sin, it was supposed, impaired our chances to achieve moral and intellectual goodness, and thus salvation. Augustine claimed that humans are fundamentally corrupt for this reason, whereas for Pelagius, "every person's sin was one's own. . . . The trouble was habit, and not nature" (Frend, 1984, p. 674). For the historical Pelagius, moral goodness and responsibility were difficult but nonetheless possible for human beings.

As a metaphor that orients our narrative of SRM, it is not the actual historical position of Pelagius that concerns us but the "popular imagination" of Pelagianism within the western Church. Pelagianism went against the orthodoxy of Augustine. For Augustine, the human soul is so corrupt that it is unable to work toward salvation without the prevenient grace of God. Pelagianism, in contrast, focuses on the human soul as free from original sin and corruption—and therefore, humans can make good choices related to their redemption:

> . . . the heresy "Pelagianism" has generally been drawn up in reference to Augustine's theology of grace. Pelagius believed that God gave grace to human beings, certainly, but this primary grace was the freedom to choose and respond. Those who chose the path of goodness would be given further encouragement by God to progress in the spiritual life. Augustine believed that such a view would render Christianity into a simplistic cult of moral "self-improvement." (McGluckin, 2004, p. 257)

Pelagianism has been a term used to describe the assumption that humans work toward their own salvation and have the capabilities and gifts necessary to do so without requiring God's grace. Through exertion we can overcome our failings and past wrongs, the popular image of Pelagianism seems to say. Pelagianism was combated by Augustine, but in many ways it continues to be debated. For example, Arminianism, a movement within early modern Dutch Calvinism, was accused of being "semi-Pelagian" because of its views on grace and predestination. More recently, we see Pelagian undertones in the popular "prosperity gospel," which preaches wealth and health as a reward for faith.

The reason such theological views continually emerge is quite understandable: humans are defined by Christianity as created "in the image and likeness of God" (following Genesis 1:26). Such a position places humanity at lofty heights, associating human reason or other qualities with divinity. This provides a justification for human uniqueness, as well as for the confidence in human ingenuity. Humans are "created co-creators"—"the human being is created by God to be a co-creator in the creation that God has brought into being and for which God has purposes" (Hefner, 1993, p. 35). And if humans are stewards of the world, science and technology is the work of God as carried out by humanity. Such "human exceptionalism" reached its heights during the Enlightenment in ways that extended beyond religion. Anna Petersen notes "secular modernists, especially after Descartes, transformed the Reformation's emphasis on the individual's solitude before God into a doctrine of self-sufficient individualism that needed not even God for fulfillment" (2001, p. 39). If such religious and philosophical considerations do not suggest a more optimistic assessment of human capability, then psychology warrants it: "there is evidence from psychotherapy that too negative a view of human nature and too low an estimate of ourselves can be harmful. Guilt without forgiveness or self-hatred without self-acceptance seem to hinder rather than encourage love of others. Some theologians join psychologists in calling for a self-respect that is not self-absorption" (Barbour, 2002, p. 53). The term *Pelagianism* expresses a theological emphasis on the boundless creativity and will of the human intellect as it engages the world.

There are similarities between the optimism of the human spirit expressed by Pelagianism and the theological framework implicit in geoengineering proposals. Both assume that the human intellect and spirit are not entirely corrupt but instead have a fundamental freedom. Thus both suggest the presence of an intellectual capacity to work toward our salvation—and it is our prerogative to do so—whether it is a salvation from sin or the worst effects of climate change. There is a balance within the historical Pelagian view: moral choice is difficult, and it is hard labor to make the correct choices. Yet humans *do* have a kernel of inborn grace, and our ability must be acknowledged. Such a balance can be found in works advocating research into SRM, even by proponents such as David Keith et al.:

> It is a healthy sign that a common first response to geoengineering is revulsion. It suggests that we have learned something from past instances of over-eager technological optimism and subsequent failures. But we must also avoid over-interpreting this past experience. Response management of climate risks requires sharp emissions cuts and clear-eyed research and assessment of SRM capability. The two are not in opposition. (2010, p. 427)

These kinds of assessments imply that humans can make the hard choices and fulfill the difficult tasks ahead of them. Human technological achievement is a matter of capability and clear vision. Our abilities provide a grace that might lead to salvation in the midst of anthropogenic climate change.

Such optimism is not necessarily created from out of an explicitly theological definition of the human condition. It can also emerge from a pragmatic appraisal of scientific method. In a provocative and influential editorial published in 2006, Paul J. Crutzen suggested that the policy impasse on climate change made the study of albedo enhancement necessary. In a reply to Crutzen, Jeffrey Kiehl wrote, "[a] basic assumption to this approach is that we, humans, understand the Earth system sufficiently to modify it and 'know' how the system will respond" (2006, p. 227). Kiehl's comment makes plain one of the main presuppositions related to all SRM proposals: humans have the prerequisite skills to mold global and regional climates and we can deal with unintended or severe consequences. In other words, if we "put our minds to the task," we are able to solve the scientific and engineering problems that geoengineering presents. Framed by the Pelagian narrative, geoengineering is not undertaken out of pride but as a true desire for "the lesser of two evils." That is to say, it can be interpreted to be an effort at using human creativity and knowledge in "a serious attempt to make amends" and "to lessen human suffering, to protect non-human species, and to preserve environmental values. . . ." (Preston, 2011, p. 470). In this case, SRM is not artificial, but *natural*: it is an outgrowth of the natural human capacity toward reason, justice, and benign dominion.

Furthermore, SRM can align with a theological pathway to salvation, not condemnation. The "salvation" of SRM relies on a view of scientific knowledge that depicts the human relationship with the world in light of a unique confidence and ultimate concern. Technology and engineering are simply manifestations of the human participation in this salvific journey. Ralph Cicerone, as part of his response to Crutzen's editorial, suggests how geoengineering research should be undertaken:

> We should proceed as we would on any other scientific problem, at least for theoretical and modeling studies. First, the underlying concepts should be identified and described. Then one should develop one or more mathematical models based on scientific principles and mechanisms. . . . Side effects that

can be anticipated should be analyzed and unanticipated side effects should be sought. Any irreversible feature of the intervention or its consequences should be identified. (2006, p. 223)

In this description, there is an assumption that science is inherently progressive and furthermore that human knowing moves inevitably toward a greater precision in its modeling of the truth. At the same time, Cicerone advocates a moratorium on large-scale projects until a careful process of judging projects—including perhaps an international body responsible for vetting proposals—is in place. But this process is built upon a certain perception of humanity and the place of the human in the world: humans—through hard work, a good will, and careful study—can understand the world and thus master it. Cicerone concludes, "Research is needed to reduce ignorance, and it is likely that gaining an acceptable amount of knowledge before intervention will take many years. Freedom of inquiry itself has moral value" (2006, p. 224).

Seeing SRM proposals as theologically akin to Pelagianism highlights a fundamental optimism and vision concerning scientific progress. The undesirable consequences of industrialization and technology do not denounce the progressive mastery of nature. Instead, the human condition is one wherein past failures do not condemn our present abilities. Like Pelagianism, the claim of SRM is that humans—through diligence, humility, and our innate capacities—can overcome wrongdoings and "be saved."

THEOLOGY AND SOLAR RADIATION MANAGEMENT

The models provide the basis for characterizing two possible Christian religious responses to SRM and geoengineering. Individuals and Christian communities might find themselves between these two poles, depending on how closely they are aligned with particular views on soteriology and ethics, theological anthropology, and eschatology. (As noted earlier, while the present argument presents Christianity as an example, I believe that this continuum will be present in some form in most religious traditions.) I conclude by explaining these three areas and how they appear to frame a religious discussion of SRM.

The two models described above provide a way to characterize how social decisions related to SRM engage Christian theological commitments. On one side, there is a theological model (symbolized by the Tower of Babel) which suggests that using technology to overstep our human bounds—that is, to attempt to become Godlike—will only result in failure. On the other side, there is a theological model (symbolized by Pelagianism) that notes our limits but poses the possibility of "salvation" through our innate, natural abilities, and efforts. Recognizing that these positions are poles of a continuum, we nonetheless see that both of these suggest an interpretation of geoengineering and SRM that relies upon *soteriology*, the theological term for an understanding of the means and end state of salvation. Salvation often informs our actions, and therefore soteriology is related (in this case, at least) to ethics.

Each model relates SRM, salvation, and ethical responsibility in different ways. The former model (symbolized by the Tower of Babel) presents a theological rationale to condemn any attempt to employ the SRM schemes. One need only look to proposals such as Roger Angel's famous "cloud of small spacecraft" (2006), which discussed placing a cloud of mirrors between the sun and earth to understand why we might consider this akin to the Tower of Babel. The editors of the *Times* of London succinctly explain why a theological vision leads to an ethical position:

> But doesn't the sheer scale of these endeavours make their proponents wary? The adverse effects could be cataclysmic, and to foresee them all would require, well, the mind of God—which is why we suggest control of the weather be left primarily up to him. Before tinkering too much with the earth's climate system, scientists would do well to remember that while God did give man dominion over all the earth, it was the serpent who said, "and you shall be like God." (*First Things*, 2009, p. 70)

Meanwhile, the latter model (symbolized by Pelagius) suggests a different ethical connection between religion and SRM proposals: using our natural capacities, we should act in the face of difficult circumstances. Keith, Parson, and Morgan write, "The idea of deliberately manipulating Earth's energy balance to offset human-driven climate change strikes many as dangerous hubris.... We think that the risks of not doing [SRM] research outweigh the risks of doing it. SRM may be the only human response that can fend off rapid and high-consequence climate impacts" (2010, p. 426). As we have seen above, this is tantamount to a theological claim about whether action condemns or saves. The resulting moral claim on the act of scientific research emerges from this implicit theology.

As should be evident, these soteriological and ethical claims rest on another facet of these two models: the specific claims of theological anthropology each pole makes. *Theological anthropology* denotes how religious communities regard the human condition, including the attributes, powers, limitations, and ends of human existence. The reason that these models are contrary to each other is that they make fundamentally different claims about the human condition, human meaning, and the nature of human life. In other words, both have fundamentally different positions on the topics of theological anthropology. The former model, emphasizing human fallibility, holds a pessimistic view of the human condition. Dale Jamieson summarizes this view when he notes how some "believe that the 'hand of man' is implicated in most of our environmental problems and they see geoengineering as more of the same" (1996, p. 323). But the latter, more optimistic model is oriented toward human capability. In both cases, the model seeks to balance human finitude and human transcendence, but each comes to a different resolution.

This balance shows the close interconnection between salvation and the human condition, soteriology and anthropology: these theological models of SRM are not simply interested in our ability to make technical choices but also in whether SRM is a "path to salvation." Certainly this can be taken to mean a salvation from climate change and environmental catastrophes more generally. But even more, is SRM part of a pathway to a more humane, enlightened, and spiritual future—one filled with the right use of technology and living within our means? Or is SRM turning away from such visions toward a dystopic and ruined existence—an earthly damnation of our own making? For religious communities (and, of course, secular society as well), these questions might very well point to the core issues raised by SRM.

This raises a final point: a discourse of fear versus hope runs through both theology and SRM proposals. Theologically, this means that SRM raises questions of *eschatology* (the technical term for the issues surrounding the theological perception of the "end times"). Scholars have noted the discussions of fear and climate change, expressed as "climate catastrophe" (e.g., Dörres, 2010) or "climate apocalypse" (e.g., Skrimshire, 2010). The discussion of climate change is often seen in apocalyptic terms, with a fatalistic tone usually reserved for the "end times." SRM proposals stress a degree of urgency, promoting research as a way of alleviating this climate catastrophe. Yet they do so with an assumption that an even more all-encompassing, permanent fix is needed. There is a hopefulness—a hopefulness that might arise from a common religious sentiment that "hope is all that is left." In other words, SRM proposals rest on a curious tension of fear and hope. On the one hand, human-caused catastrophe is imminent, yet a human-constructed solution is possible. On the other hand, such human-engineered salvation is another outworking of the natural process that our science compromised. Christianity and other religions are also concerned with the balance between hope and fear, forgiveness and sin, good and evil. And religion, too, has a complicated understanding of hope and fear. Salvation is intertwined with the uniqueness of human reason and its inevitable failure. It is also centered on divine grace—but this is balanced by divine justice and the fractured world that results from divine retribution. This finally connects religion and climate. After all, Mike Hulme notes that "[e]xperiences of extreme weather have long been interpreted by individuals and cultures as signifiers of divine blessing or judgement" (2008, p. 7). We might surmise that whether SRM is judged favorably or not has much to do with our sense of hope in the face of a fearsome apocalypse.

By using Christianity as an example, we can see that our theological interpretations of the world and of the human being influence how we ethically assess and respond to climate change. This leads to three areas in need of further reflection. First, much work can (and should) be done in analyzing how other traditions offer a qualified acceptance or condemnation of SRM from within their own belief and ritual structures. Second, the different positions of ethics, anthropology, and eschatology represented by the two models show how our responses to SRM must find ways to balance the models of the Tower of Babel and Pelagianism. Finally, based on the discussion and sources mentioned here it might also be evident that the scientific and technological communities involved in studying and (potentially) enacting geoengineering proposals carry with them certain implicit theological frameworks. Scientists, too, have ultimate concerns; such concerns rest in a discussion of control and hubris, hope and fear, capability and fallibility, knowledge and ignorance. These frameworks help facilitate a broader understanding of the complexities of the ethics of geoengineering.

NOTES

1. These examples are related to Christianity, but there are similar discussions in other religious traditions. For a snapshot, see the major book series published by Harvard University Press on behalf of the Forum for Religion and Ecology, *Religions of the World and Ecology*. Each volume approaches an individual religious tradition (with the exception of the volume devoted to indigenous traditions) and its relationship to the environment.

2. As Pugliese and Ray pointed out in 2009, Gallup has found that the majority of the world's adult population is aware of climate change (61percent overall, 82 percent in the Americas, and 88 percent in Europe). Thus, in the last several years it has become increasingly clear the extent to which human action is a cause, why climate change is a threat to human and non-human communities, and the uncertainties that still exist in our scientific understanding. Even so, only with the awarding of the 2007 Nobel Peace Prize to Al Gore and the Intergovernmental Panel on Climate Change was global warming cemented in the popular imagination. And in the months before the Copenhagen UN meeting in 2009, there was " . . . a greater sense of urgency than in the past among many global climate scientists and experts, who say climate change is taking place faster than they anticipated" (p. 64).

3. For example, book length treatments include McFague, 2008; Northcott, 2007; Robb, 2010. In addition, Sigurd Bergmann, a theologian at Norwegian University of Science and Technology, Trondheim, has engaged in several projects related to climate change and religion. Bergmann's projects are resulting in edited collections of essays, some of which are explicitly theological in orientation.

Chapter Thirteen

Making Climates

Solar Radiation Management and the Ethics of Fabrication[1]

Maialen Galarraga and Bronislaw Szerszynski

In light of the increasing probability of dangerous climate change, a range of different geoengineering techniques—large scale, deliberate interventions into climate processes and systems—are being proposed as ways of preventing excessive warming of the global climate. Amongst these are various solar radiation management (SRM) approaches to geoengineering: techniques that would work by altering the albedo (reflectivity) of the earth and thus modulating the energy budget between earth and space. There have been proposals for applying such techniques at the earth's surface (such as painting roofs white, covering deserts with plastic, cultivating more reflective varieties of crops), in the troposphere (cloud-albedo enhancement), in the stratosphere (particle injection), and in earth's orbit (sun shields) (Royal Society, 2009).

Such technological interventions would, in an important sense, be making the climate. The very definition of geoengineering means that it is intentional and planned; the full-scale implementation of SRM would thus result in a climate that was an artifact—a climate that has not just been *disturbed* by human intervention but has been *intentionally shaped* by human intervention. For the first time, we would have a made climate, and this provokes significant questions. As many other contributions to this volume argue, SRM clearly raises ethical issues about whether such interventions should go ahead and how the risks and benefits could be shared justly (see also Jamieson, 1996; Corner & Pidgeon, 2010; Gardiner, 2011b). But in this chapter we approach the normative dimensions of SRM in a rather different way: by reflecting on what exactly it would mean for humanity to "make" the climate, and how this might draw us into a new relation with nature.

We thus construe the normative in a much broader way than is usual. Drawing on the philosophy and anthropology of technology, we approach the act of human making, what Aristotle called *poiesis*, as something that has to be grasped as a whole rather than decomposed into, on the one hand, a set of technical questions to be answered by scientists and engineers and, on the other, a set of ethical and political questions to be answered by philosophers, moralists, and politicians. We treat the bringing of new things into being not simply as one kind of activity that human beings may or may not engage in but as something that shapes and conditions the kind of beings that humans are in the world, not as an addition to an original, natural, non-technical humanity but as in some sense constitutive of the human. We suggest that any specific normative judgments about what we as humans do with our powers to make things ought to be situated against this background of how we understand the nature of *poiesis*, of human making.

More specifically, we argue that debates around geoengineering research have been shaped by one particular imaginary of making—one specific understanding of what it is to make something—which has had the unhelpful consequence of occluding crucial dimensions of what it would mean to make the climate. In order to develop this argument, we first focus on the nature of artifacts, distinguishing two ways in which something can be understood as "made," and we use this distinction to think about what kind of entity a made climate would be. We suggest that making is usually understood as "matter taking form," but we then present three very different accounts of how the en-forming of matter takes place, accounts that we call "producing," "educing," and "creating," respectively. Finally, we show how each of these models of making implies a particular kind of maker, a particular version of the human agent, and we explore how these three different visions of the human as maker can help illuminate the ways of making that are currently being enacted in the field of geoengineering—and ways in which this might be reconceptualized.

MADE THINGS

There are many different ways that artifacts could be categorised, but most pertinent to our argument is a distinction that can be made in the way that different kinds of artifacts persist over time. Made things, like other entities, do not all endure in the same way. Here we want to distinguish "stable" from "metastable" artifacts. *Stable artifacts*, such as tables and chairs, endure in a way that depends on minimizing the exchange of material and energy across their boundaries. For example, when making a table, the wood is typically planed, sanded, and varnished, in order to prevent bits of wood from breaking off the table and water from penetrating it. This is because the natural tendency of a stable artifact is to persist, but to the extent that its boundedness is not maintained, it slowly degrades by losing order and separation from its environment. In the language of Gilles Deleuze, with this kind of artifact, it is the *extensive* properties such as length, volume, and mass—the properties that together comprise its stable, completed form—that are most important. *Intensive* differences such as temperature, pressure, and density are either seen as accidental and not constitutive of it being the kind of artifact that it is or, if they are essential to its operation, they are typically seen as subordinate to and derivative of the artifact's extensive properties—for example, the power and internal dynamics of an internal combustion engine are seen as a result of its size and shape, not vice versa.

With this kind of artifact, the act of fabrication is completed when the artifact has been made: in Aristotle's (1956) terms, this kind of making is a *kinesis*, an instrumental action, one whose goal lies outside of it. This kind of fabrication is thus a self-destroying process in that when the artifact is made and made stable, the process thereby loses its purpose and naturally finishes (Arendt, 1958, p. 143). In such cases, if the action is continued or repeated after completion, this must be for reasons external to the process. For example: once a craftsperson has finished making a table, he or she might make another table, but this would not be because the making of a table requires it, but perhaps because the table-making is embedded in a larger structure of action such as artisanal labor (Arendt, 1958, p. 143).

By contrast, *metastable artifacts*, such as fires, fields, and gardens, are artifacts that maintain their existence dynamically through the controlled exchange of material and energy with their environment. This kind of artifact is a subset of what the Nobel prize-winning chemist Ilya Prigogine called "dissipative structures" (Prigogine & Glandsorff, 1971). Dissipative structures—such as living organisms, ecosystems, or societies but also abiotic structures such as cyclones, fires, hurricanes, and convection cells—are open systems that are far from equilibrium, so they exchange energy and/or matter with their environment as described by the second law of thermodynamics. However, *because* they are far from equilibrium, non-linear rather than linear dynamics are able to dominate, and this allows processes of self-organization to arise (Shao et al., 2002, p. 57). Such structures can thereby avoid moving to thermodynamic equilibrium and thus actively maintain their existence—not just despite, but because of their dissipative form—by exporting entropy to their surroundings. Certain kinds of flows of energy and matter thus take place. Rather than canceling out intensive differences, these flows maintain them (De Landa, 2005, p. 82; see also Bogue, 1989, pp. 61–62). These are self-organizing systems in which the form evolves out of the interactions of matter; in contrast to the stable artifacts discussed above, the extensive properties of metastable artifacts are the result of the ongoing play of intensive differences.

An artifact that has this kind of mode of persistence might exhibit a range of different dynamics over time. It may actively maintain or even increase its internal order and distinctiveness from its environment. It may have its own inherent tendency to stability due to the existence of "attractors," "singularities," or "basins of attraction"—more-or-less stable dynamic states to which it naturally evolves over time. It might be extinguished suddenly and catastrophically. But it might also persist, becoming progressively less artifactual if the metabolic exchange across its boundary is not continually modulated. For example, a campfire in a forest can become a forest fire. This kind of artifact thus requires a continuous action of tending, cultivating, and caring to maintain it in its artifactual state. With this kind of artifact, it is thus inherent that the making is never complete—in Aristotle's (1956) terms, it is an *energeia*, an ongoing kind of action that never reaches a natural terminus. As Hannah Arendt says of such artifacts, "[a] true reification . . . , in which the produced thing in its existence is secured once and for all, has never come to pass; it needs to be reproduced again and again in order to remain within the human world at all" (1958, p. 139).

We will make use of this distinction later in this chapter. The "natural"—or at least non-geoengineered—global climate is clearly a metastable rather than a stable entity: the (extensive) shape of the atmosphere's complex dynamic structure of atmospheric cells, wind belts, cyclones, anti-cyclones, and so on is a dynamic product of the (intensive) dissipation of energy from the equator to the poles. We should therefore ask whether a climate made through SRM would necessarily be a metastable artifact, one that would need to be constantly tended and modulated. But we should also ask about how the form of a made climate would emerge from our interactions with climate processes. In the next section we thus move from the "made" to "making"—from thinking about what a "finished" artifact is to exploring different accounts of what it is to make something.

MAKING THINGS

In the previous section we have already distinguished between two forms of making: the time-delimited making that terminates in the production of stable artifacts and the ongoing practices of cultivation and tending required to maintain the form of metastable ones. But there are other distinctions that we need to make before we look at the making of climates, distinctions that will start to complicate this contrast. In this section we thus draw on a number of philosophers to develop three different accounts of what it is to make something, which respectively we call *production* (imposing existing forms onto matter), *eduction* (drawing forms out of the potentiality of matter), and *creation* (creating radically new forms by rearrangements of matter)—all of which are relevant to thinking about what it might mean to make climates. In one sense, they are describing different kinds of processes used in the making of things, in another, they are competing accounts of what it is to make something—"imaginaries"—each of which can lead us to conceive of making in a particular way, guiding the manner in which making proceeds and constraining our awareness of what is at stake when something is made.

All three of these understandings of making could be described as "hylomorphic" in the sense that they have their origins in Aristotle's understanding of making as involving some kind of dynamic relationship between matter and form.[2] In the *Physics*, Aristotle (1929) develops an understanding of objects, whether natural or artificial, as a compound of two metaphysically distinct elements: matter (*hyle*) and form (*morphe* or *eidos*). Matter for Aristotle was anything from which a thing is composed, whether material or immaterial, animate or inanimate. For Aristotle, form is actuality—the shape that a thing needs to have to be what it is (for example, a bronze statue is only a statue by virtue of its form; without that form, it is simply bronze). Matter, by contrast, is pure potentiality—it has the potential of becoming different things (for example, clay is a potential bowl and also a potential brick; it only actually becomes one of these things when suitably en-formed). But as we shall see, although the three different ideas of making that we shall consider use the language of form and matter, they understand the nature of making in very different ways.

Our first model of making is *production*, understood here as the imposition of an existing form onto formless matter. The clearest cases of production in this sense would be activities like using a mold to give a form to setting clay or dough or like using a machine press to cut and shape sheet metal, but a wide range of other fabrication processes are conceptualized and organized in these terms. The first thing to emphasize about this model of making is the passive role that it ascribes to matter. Aristotle follows Plato in arguing that in the act of fabrication, reason is guided by *eidos* or form: Plato had argued that the craftsman making or repairing a weaving shuttle, like the demiurge who made the universe from formless matter, allows his hands to be guided not by the matter (*hyle*) out of which the artifact is to be made but by the eternal, unchanging form (*eidos*) of the finished product. Hannah Arendt also follows this understanding of fabrication as guided by form—by a model, whether beheld in "the eye of the mind" or in a physical plan (1958, pp. 140–41). Fabrication for Arendt involves a relation of domination toward nature and matter; even the material worked on by the fabricator has had to be wrested from nature. Heidegger suggests a similar view when he argues that, in the production of equipment of useful things such as a stone axe, "stone is used, and used up. It disappears into usefulness. The material is all the better and more suitable the less it resists perishing in the equipmental being of the equipment" (1977, p. 171).

Our second account of making artifacts is what we call *eduction*. Whereas production involves the imposition of form onto matter from outside, in eduction, form is drawn out from the potentialities of matter itself.[3] This model of making thus has a very different understanding of how a stable form emerges, one that is closer to the tending and cultivating of metastable artifacts. In Deleuze's terms, whereas production involves "commanding matter," eduction requires one to "surrender" to it, so as to be able to coax a physical system toward one of its "singularities," a threshold beyond which self-ordering takes place (Protevi, 2001, p. 9). Ingold describes this as a difference between "architectonic" and "textile" making (Ingold, 2010b, p. 92–93). According to Ingold, eduction involves a kind of "weaving," a following of the inclinations of matter and an intervening in fields of force and flows of materials in order to shape the way that they unfold and stabilize. Whereas making-as-production prioritizes the final form, making-as-eduction focuses on the process of formation itself and sees any final, stable form merely as the frozen last episode of a series of transformations of material (Mackenzie, 2002, p. 47; Ingold, 2010a, p. 3). Thus, whereas production focuses on the *extensive* characteristics of the resulting, actualized form, eduction attends more to the role of the *intensive* potentialities involved in the bringing of forms into being.

Theorists of eduction emphasize that even the making of identical, stable artifacts with a form that is planned in advance might in practice be closer to eduction than production in character. Thus, in Gilbert Simondon's account of making a brick from clay and mold, he argues that it is not the case that the clay is pure "matter" and the rectangular mold pure "form," or that the mold actively impresses the form onto the passive clay. In order to be able to take the form of a brick, clay has to be in a state of metastability—apparently stable and inert but full of potentialities. The clay's capacity to take a form and its capacity to hold a form are one and the same. The mold does not impose a form; it rather provokes an "internal resonance" that transforms the potentialities in the clay into a determinate equilibrium. Similarly, the form-making does not just take place at the surfaces where the clay touches the mold, but all the way through the ensemble of the tamped clay and the tensioned mold around it (Simondon, 1964, pp. 37–43). Furthermore, the distinction between a finished, enduring artifact and one that has to be continually cultured and tended may be illusory because of the need for "ongoing processes of formation" (Mackenzie, 2002, p. 49). For example, not only the artisans who make artifacts but also the users of artifacts constantly have to improvise, shore up, and repair to allow the artifact to endure. As Ingold remarks, "Like life itself, a real house is always work in progress and the best that inhabitants can do is to steer it in the desired direction" (Ingold, 2010b, p. 94).

Our third model of making is *creation*. Whereas with "eduction," the distinction from "production" was the former's emphasis on the autonomy of *matter*, i.e., its potentiality to suggest form, with creation the emphasis is on the autonomy of *form*, i.e., the capacity to create a new *eidos* that is not wholly determined by anything that pre-exists it. Here we are mainly drawing on the work of Cornelius Castoriadis, who emphasizes the capacity of human beings to bring into being radically new forms. Castoriadis argues that producing artifacts in the sense discussed above is really not making anything new because it is simply impressing existing forms onto existing matter:

> If we imprint upon a mass of bronze an *eidos* that is already given, we are merely repeating what essentially, as an essence—*eidos*—was already there, we are creating nothing, we are imitating, we are *producing*. Conversely, if we make *another eidos*, we are doing more than 'producing', we are *creating*. (1987, p. 197)

Once a new kind of thing is brought into being, the further reproduction of its form in material form will constitute "production"—the proliferation of instances, copies of, or variations on the original artifact (1987, p. 180). But the initial act of creating a new "type"—a tool (such as a knife or an adze), an institution (such as a temple or a school), or a social phenomenon (such as inheritance or an election)—is an "ontological genesis": the emergence of radical novelty, which can be understood neither as the imposition of an existing form onto matter nor as the mere drawing out of a form that was already lying latent in matter. Creating is *ex nihilo* (out of nothing), and thus implies a departure from the Platonic paradigm of a timeless world of ideas (*eide*) that represents everything that could ever be. Understanding making as creation also requires us to understand time as genuinely historical, as an irreversible sequence whereby what comes into being (the created *eidos*) is unique and pertains to a specific point in history.

The notion of making as creation challenges traditional western ideas of matter as "ready given," stable, or as simply discovered and made use of by living beings. Life (the living being) does not encounter a ready-given "*world-out-there*" but creates it and in doing so, also posits itself *as* a living being. As Suzi Adams puts it, "Prior to the emergence of the living being there is neither 'world' nor 'meaning' (Adams, 2007, p. 85). Castoriadis differentiates a number of levels in this self-creation, from the living being up to the emergence of human society (Adams 2007), and it is at the level of the human that this theory interests us more here. According to Castoriadis, each new form of society is the emergence of a new *eidos*, the "opening up of a world," a radically new way of organizing thought and action. For example, the notion of "individual" that arises during the European Renaissance cannot simply be derived from the concept of the human being in the same way as the calf is derived from the cow—it adds something new: in that sense it is an invention of another version of the living being rather than a mere reproduction. For Castoriadis, acknowledging this fundamental self-creative capacity of ourselves *qua* humans constitutes us as autonomous entities and enables us to put into question (and consequently re-build accordingly) the world that we have created (Adams, 2007; Hansen, 2009).

THE MAKER OF CLIMATE

In the previous section, we looked at three different idealized accounts of the ways in which artifacts might be made, accounts which differed in their understanding of the relationship between form and matter in the act of making: production (the imposition of existing forms onto matter); eduction (modulating intensive differences to create stable, enduring artifacts); and creation (bringing radically new forms into being and thus creating a "world"). In this section we will explore how these different models of making each imply a different vision of the maker of things and how these different visions of the maker might shape the understanding and practice of climate engineering through SRM. We propose three archetypal makers, corresponding to production, eduction, and creation, respectively. We name these the *architect*, the agent of production who dictates in advance and from up on high what is going to be produced; the *artisan*, who facilitates the potentialities of the material on which he or she is working in a constant process of formation; and the *artist*, who brings something radically new into existence, and thereby creates a world.

The *climate architect* is the agent who it is imagined can impose a new form onto the matter of climate. We develop this term from Paul Protevi's commentary on the work of Simondon and Deleuze, in which the "architect" is a figure imagined to command matter from a distance, who, in Simondon's words, "imposes a form onto a passive and indeterminate matter" (Simondon 1964, p. 48-49; see also Protevi, 2001, p. 8). This is the picture of the maker of climate that currently dominates the contemporary discourse of geoengineering, one which involves, firstly, forming a predefined idea of a possible climate to be achieved and, secondly, actualizing that form by somehow impressing it onto the matter of climate. The climate architect is an idealized, imagined figure who knows in advance the form that they want the climate to take, who can identify the process whereby they can provoke the climate to take it, and who can carry out that process and bring the matter of climate into the desired form. If uncertainties are acknowledged in this way of thinking about making the climate, they are seen as factors that are exogenous to the process of production itself and as in principle capable of being ironed out by future technical refinements.

This "productive" way of thinking about making climates is arguably encouraged by the central role played by computer models in climate science, including geoengineering research. Despite the awareness of individual climate scientists about the limits of their computer models, there is still a tendency to treat climate models as "truth machines" that can reveal the actual form of future climates under various mitigation and geoengineering scenarios (Wynne, 2010). In SRM research, for example, computer-based simulations of specific interventions are used to generate representations of possible future climates in the form of tables, graphs, and maps. This kind of scientific practice has the effect of rendering climate as pure information *in silico*—as form stripped of matter. This dematerialized, formal climate can then be imagined as something that can be recombined with matter and thus made actual.

But imagining a climate architect is to imagine a world in which the "matter" of made climates is "fixed, stable, and uneventful" (Mackenzie, 2002, p. 40). But even if a made climate were continually modulated, would it really be possible to impress a predetermined form onto the metastable climate system? To use Heidegger's language, could the matter of climate be made to "disappear into usefulness," to give its being up to a form chosen by us to serve the well being of humanity? In the case of making a brick, the process of en-forming the clay with the help of the mold is dependent on the clay having been purified—for example, by removing any clots or stones that would act as "parasitic singularities" and disrupt its en-forming (Simondon 1989, p. 42). Yet the climate cannot be purified in this way. Like an organism, the climate system is continually in formation, with constant adjustments of flows of energy and matter, temperature inversions, and boundaries between pressure systems. What happens in an instant and more-or-less irreversibly in the making of a brick happens continuously in a metastable entity like the climate system as inherent incompatibilities—tensions between different intensive states—are continually resolved through processes of internal resonance and exchange of energy (Mackenzie, 2002, p. 50). This suggests that the uncertainties in climate models are not mere "noise" that could in principle be erased but are the result of potentialities that are intrinsic to the way that the climate maintains and develops its form over time. It also suggests that a more appropriate model for the making of climates might be eduction and a more appropriate figure for the maker of climates, the artisan.

One analogy that might be useful in capturing the notion of the *climate artisan* is provided by David Turnbull's (2000; see also Ingold, 2010b) account of the practice of cathedral building in the Middle Ages. Before architecture became associated with abstract, pure geometrical forms that only exist in the mind or on paper, medieval master builders worked alongside and coordinated their masons on site, and the plan of the building emerged out of the task of making, measuring, and assembling. There was no abstract *a priori* idea of what was to be designed, no standardized, homogenous material to be carved and assembled; rather, building the cathedral meant surrendering to the medium and allowing the form to emerge out of the process of building. Similarly, the climate artisans would not seek to identify the final form of the made climate in the mind or *in silico* and then try to achieve it in matter; instead, they would allow the form to emerge out of their interactions with matter, through recursive learning and adjusting. Their focus would be less on the desired form to be taken by the climate and more on the actual process whereby the en-forming of climate might take place. They would thus treat computer models not as truth machines that can be used to reveal the future, but as laboratories or sandpits in which skills can be learned and instincts sharpened—in which the beginnings of a "feeling for climate" might be cultivated, comparable to the "feeling for the organism" described by the biologist Barbara McClintock (Fox Keller, 1983). Just as the mold in Simondon's brick-making example is, in a sense, a frozen manifestation of the molding hands of the fabricator (Simondon, 1989, p. 40), we can similarly consider the "molding hands" of the climate artisan as being virtually presenced through the mediating technologies of computer models and SRM delivery systems.[4]

However, the climate artisan model also has its limitations as a way of understanding the maker of climate. The necessity for the artisan to follow the inclinations of matter in order to educe form means that, in the case of powerful, metastable natural phenomena, it is more the case that the actions of the artisan are governed by the material to which they have harnessed themselves than the other way around. John McPhee's detailed descriptions of heroic attempts to control river floods, mountain landslides, and lava flows show how ambiguous is the phrase "the control of nature" when dealing with such phenomena (1989). When humans progressively entangle themselves in the potentialities of matter in order to modulate the dynamics of nature's becoming, "control" typically becomes at best a two-way process and one in which the intentions of the humans become conditioned by those of nature. Geoengineering by SRM is likely to be just such a scenario, one in which the climate artisan becomes tied to the continuous task of modulating climatic processes and thus subject to their logic. As James Lovelock suggests, if we start to use SRM, we are likely to be fating ourselves to a continuous process of correction and adjustment in order to counteract the unanticipated effects of each intervention, and we will quickly find ourselves "enslaved in a Kafka-like world from which there is no escape" (Lovelock, 2008, p. 3888). The climate artisan would thus be not *homo faber*, the human being as the fabricator of the enduring things of a made world, but *animal laborans*, the laboring animal who serves the endless processes of life's self-maintenance (Arendt, 1958). The movements of the climate artisan would be determined by the resistances and inclinations of matter—like the craftsman of preindustrial societies who, because of the simple tools at their disposal, had to learn and perform complex "habits, gestures and schemes of action," often working in teams, in order to complete technical operations such as shoeing a horse, building, or threshing (Simondon, 1989, pp. 77-79),

But the climate artisan model also suffers from another crucial limitation. Like the climate architect, the climate artisan imagines the final form of a made climate as in some sense pre-existing its actualization—perhaps not in the matterless form of an *in silico* modeled climate but instead as latent in the matter of climate—to be coaxed out through a series of transformations and adjustments. Neither the climate architect nor the climate artisan is oriented toward the radical novelty that made climates might entail. Similarly, like that of the climate architect, the horizon of thought of the climate artisan does not contain the place for an awareness of how making a climate could also be the making of a world. For this kind of orientation we need to move beyond both the climate architect and the climate artisan.

So finally we turn to the figure we call the *climate artist*, who approaches the making of climates as an act of creation in the sense that we are using it here. Being a climate artist would involve thinking through the idea that making climate would inevitably involve creating climatically novel states. This kind of making of climate is likely to be unavoidable because of the impossibility of simply returning the climate to its preindustrial state. For example, if it is really the case that the pattern of stratospheric aerosol injection needed to restore the temperature distribution to that of a low-CO_2 planet is not the same as the distribution needed to restore the hydrological cycle, then judgments about which climate is desirable are unavoidable (Ban-Weiss & Caldeira, 2010; Lunt et al., 2008). In such circumstances, even though geoengineering has been defined in terms of the intention to ameliorate the effects of raised CO_2 levels—as if it were the returning of the matter of climate to its preindustrial form—it would be problematic to describe SRM as climate restoration or remediation. The climate artist would understand the creation of novel climates not as an accidental failure of geoengineering interventions to accurately "counter the climate effects of past greenhouse gas emissions" (Bipartisan Policy Center, 2011, p. 3)—as if this would re-create the preindustrial climate—but as an intrinsic aspect of making the climate, the implications of which need to be reflected upon.

But a climate artist would also have to recognize climate creation, like the creation of a work of art, as world-making. The climate artist goes beyond the architect and the artisan in that (s)he approaches the making of climates not as the mere assembling of climatic processes but as the actualization of "something more" (Castoriadis, 1984, p. 234). A climate artist would realize that the creation of climate might also mean the bringing into being of a new kind of society with a new articulation of what climate is and how we relate to it. Both the climate architect and the climate artisan imagine that it is possible to maintain some kind of continuity and consistency between the goal formed in advance and the final achievement of a made climate. A climate artist, by contrast, would recognize that the creation of a new *eidos* produces a historical rupture, a new context in which ways of thinking and forming intentions can be utterly transformed. Seeing SRM as involving creation in this sense means that it could not simply be judged as a means to an end and thus as capable of being deemed successful or unsuccessful by criteria set in advance. Instead, its deployment would have "changed the end in changing the means" (Latour 2002, p. 252); it would have created a new world, a new kind of society, in which geoengineering itself is likely to take on new meanings, be put to new uses, and be judged in new ways. Discussions about what constitutes "responsibility" in geoengineering research, deployment, and governance need to engage with such challenging but crucial dimensions.

Furthermore, if geoengineering were to be the *creation* of climates, then, as atmospheric entities and processes became the matter of made climates, this would alter what they were for us. Thinking of a made climate as a "work" enables us better to grasp how creating a climate could change not only the form but the very matter of climate. Castoriadis points out that all societies have to deal with something that they treat as matter—i.e., that which resists the will but is malleable. But what *counts* as matter, how it resists, and how it is malleable is different in every society: for example, only in our society can hydrogen atoms be fused together (Castoriadis, 1987, p. 355). Similarly, in a geoengineering society, the matter of climate would become matter in a radically new way because of our very ability to manipulate it. The question, however, would be whether the made climate would be treated as "equipment"—judged by its mere utility—or as a "work"—judged by its world-making power. In the first case, the matter of climate would be expected to simply "disappear into usefulness"; in the second, it could be allowed to "shine forth" in a new way, as the matter taken up into a work of art is given a new way of presencing itself (Heidegger, 1977, p. 172). The climate artist thus has to take on another level of responsibility: to attend to the way that making climates will change what the climate, the sky, and the weather are for us—their meaning and place in human society.

CONCLUSION

In this chapter we have explored different kinds of artifacts and different models of making in order to try to open up some new questions about geoengineering techniques such as SRM. We have argued that the discussion of the science, governance, and ethics of geoengineering has been dominated, explicitly or implicitly, by a particular imaginary—a particular idea of what it is to make something: that of production, as involving the bringing together of a pre-existing form and pre-existing matter. We have suggested that the reality of SRM is likely to be inadequately captured by this way of thinking about making and that we need to consider alternative accounts of what it is to make something if we are to appreciate much of what would be at stake. We used the ideas of eduction and creation in order to suggest other ways of thinking about how the task of making an artificial climate might unfold. Of course, the three models we have discussed are ideal types—any real-world implementation of SRM is likely to combine elements of more than one of them. But we would argue that the distinctions that we have drawn between the three models allow us to identify important points of divergence between different ways that SRM might be conceived, with significant implications.

For, although they are models of making—about the practice of forming matter—this does not mean that they deal solely with the technical. We have tried to show that, embedded at the heart of apparently practical questions of how it may or may not be possible to shape the climate, there are profound normative issues. The conventional way of thinking about the ethics of technology would lead one to imagine that the science and technology of geoengineering are completely distinct from issues of ethics and governance, separated by the gulf that divides fact from value, is from ought. But the very idea of making the climate has to draw on particular models of fabrication, and the choice between these models raises metaphysical questions about the human place in the world as a fabricating being. These models do not necessarily by themselves lead to specific moral positions on the permissibility and acceptability of particular acts, the distribution of risks and benefits, or the right to be consulted. But they force us to think about what it is to be a being that makes things and what it might mean to bring the climate into the orbit of human making. What kinds of responsibility are inherent in the act of making? Does the making of climate only involve responsibility in respect to the final, made, extensive form? Or should we attend more to the responsibilities internal to the ongoing, perhaps endless, action of making the climate, conceived as a vocation, a praxis, in which climate artisans bind themselves in relations of responsiveness to the inherent inclinations of climate—its singularities and intensities? Furthermore, should making the climate be seen as also involving an awareness of the world-making power of creative acts and thus a responsibility for and to the kind of world that geoengineering might bring into being?

We have represented the climate artist as in some way the more comprehensive of the three models of the maker of climate in terms of the kinds of responsibility that it is possible to conceive within its horizon of thought. However, we should make it clear that we are not advocating any of the three roles. All of them, in different ways, are deeply disquieting prospects, and spelling them out should be enough to make clear what a serious step it would be for geoengineering to go ahead—if this needed to be made clear at all. It is not at all certain that humanity has the ability to take on any of these three roles in respect to the climate. However, the idea of the climate architect is perhaps the most worrying of the three because of the difficulty of articulating important dimensions of the ethics of making within its terms of reference.

Recent technological developments such as those in biotechnology have prompted reactions from the public that scientists are "playing God" (Davies, 2006). Other writers have embraced the language of becoming-God in a more positive way in order to try to call humanity to a greater awareness of its power and responsibility in the Anthropocene epoch (Lynas, 2011). If explorations into the feasibility and desirability of geoengineering the climate continue, we can expect such language to be drawn upon by both opponents and proponents as a way of expressing the seriousness of the step that we are considering. But another way to frame the key question that we have been pursing in this paper is to ask: What *kind* of god would we become if we started to make the climate? It would be better to reflect on such questions in advance, before we find ourselves being drawn into a role that we have not freely chosen.

NOTES

1. Earlier versions of this paper were presented at a seminar at the Institute of Advanced Study, Durham University, on November 14, 2011, and at the workshop Terra infirma: Experimenting with geo-political practices, University College London, on January 27, 2012. The authors would like to thank Christopher Preston, Adrian Mackenzie, Andy Jarvis, Kathryn Yusoff, and participants at the above events for helpful comments and suggestions. Maialen Galarraga gratefully acknowledges the support of EPSRC/NERC grant EP I014721/1, Integrated Assessment of Geoengineering Proposals (IAGP). Bronislaw Szerszynski would like to thank Lancaster University for sabbatical leave and Durham University's Institute of Advanced Study for a visiting fellowship in autumn 2011, both of which helped make writing this paper possible.

2. "Hylomorphism" is usually reserved for the kind of making we are calling "production"—the imposition of form onto matter—but we prefer to use it in a broader way that applies to all three of our models of making. Amongst theorists of what we will call "eduction," there is an attempt to get away from the language of "matter" and "form" in favor of the idea of an

active "material" with its own singularities and tendencies that is open to and captures cosmic "forces" and "intensities" (Deleuze and Guattari 1988, pp. 337–50). But for the sake of simplicity we are remaining with the language of form and matter in describing "eduction."

3. The term "eduction," like "education," is derived from the Latin *educere* "to lead out, bring out," itself derived from *ducere*, "to lead," the root of "produce," "reduce," "deduce," "induce," and so on.

4. Some elements of a climate artisan approach are explored by Lempert and Schlesinger (2001) and Jarvis et al. (2009).

Bibliography

Adams, S. (2007). Castoriadis and autopoiesis. *Thesis Eleven, 88*(1), 76–91.
Adams, S. (2008). Towards the post-phenomenology of life: Castoriadis' critical naturphilosophie. *Cosmos and History: The Journal of Natural and Social Philosophy, 4*(1–2), 387–400.
Agrawala, S., Ota, T., Ahmed, A. U., Smith, J., & van Aalst, M. (2003). Development and climate change in Bangladesh: Focus on coastal flooding and the Sundarbans. Retrieved from http://www.pisa.oecd.org/dataoecd/46/55/21055658.pdf
Allenby, B. (2010). Climate change negotiations and geoengineering: Is this really the best we can do? *Environmental Quality Management, 20*(2), 1–16.
American Physical Society. (2011). *Direct air capture of CO_2 with chemicals: A technology assessment for the APS panel on public affairs.* Retrieved from http://www.aps.org/policy/reports/assessments/upload/dac2011.pdf
Amsler, S. S. (2010). Bringing hope "to crisis": Crisis thinking, ethical action and social change. In S. Skrimshire (Ed.), *Future Ethics: Climate Change and Apocalyptic Imagination* (pp. 129–152). New York, NY: Continuum.
Anaya, S. J. (2004). *Indigenous peoples in international law* (2nd ed.). New York, NY: Oxford University Press.
Angel, R. (2006). Feasibility of cooling the earth with a cloud of small spacecraft near the inner Lagrange point (L1). *Proceedings of the National Academy of Sciences, 103*(46), 17184–17189.
Arendt, H. (1958). *The human condition.* Chicago, IL: University of Chicago Press.
Arendt, H. (1961). *Between past and future.* New York, NY: Viking Press.
Aristotle. (1929). *Physics (2 Vols.)* (P. H. Wicksteed & F. M. Cornford, Trans.). London, England: Heinemann.
Aristotle. (1956). *Metaphysics* (J. Warrington, Trans.). London, England: J. M. Dent & Sons.
Arrow, K. J. (1963). Uncertainty and the welfare economics of medical care. *American Economic Review, 58*, 941–973.
Arrow, K. J. (1965). *Aspects of the theory of risk–bearing.* Helsinki, Finland: Yrjö Hahnsson Foundation.
Arrow, K. J. (1968). The economics of moral hazard: Further comment. *American Economic Review, 58*, 537–539.
Arrow, K. J. (1985). The economics of agency. In R. J. Zeckhauser & J. W. Pratt (Eds.), *Principals and Agents: The Structure of Business.* Boston, MA: Harvard Business School Press.
Asilomar Scientific Organizing Committee. (2010). The Asilomar Conference recommendations on principles for research into climate engineering techniques. Retrieved from http://climateresponsefund.org/images/Conference/finalfinalreport.pdf

Baillie, J. E. M., Bennun, L. A., Brooks, T. M., Butchart, S. H. M., Chanson, J. S., Cokeliss, Z., Stuart, S. N. (2004). A global species assessment. Retrieved from http://data.iucn.org/dbtwwpd/html/Red%20List%202004/completed/cover.html

Bala, G. (2011). Counteracting climate change via solar radiation management. *Current Science, 101*, 1418–1421.

Bala, G., Caldeira, K., Nemani, R., Long, C., Ban–Weiss, G. A., & Shin, H. J. (2010). Albedo enhancement of marine clouds to counteract global warming: Impacts on the hydrological cycle. *Climate Dynamics, 37*(5), 915–931.

Ban–Weiss, G. A. & Caldeira, K. (2010). Geoengineering as an optimization problem. *Environmental Research Letters, 5*(3), 1–9.

Banerjee, B. (2011). The limitations of geoengineering governance in a world of uncertainty. *Stanford Journal of Law, Science & Policy, May*, 15–36.

Barber, K. (2005). *Death of Celilo Falls*. Seattle, WA: University of Washington Press.

Barbour, I. (2002). *Nature, human nature, and god*. Minneapolis, MN: Fortress Press.

Barker, O. (2009, August 4). Got that virtual glow? Texting, tweeting render users "present yet absen." *USA Today*, pp. 8B–9B.

Barnett, J. & Adger, W. N. (2007). Climate change, human security, and violent conflict. *Political Geography, 26*, 639–655.

Barrett, S. (2008). The incredible economics of geoengineering. *Environmental & Resource Economics, 39*(1), 45–54.

Baumert, K.A., Herzog, T., & Pershing, J. (2005). Navigating the numbers: Greenhouse gas data and international climate policy. Retrieved from http://pdf.wri.org/navigating_numbers.pdf

Benatar, S. & Singer, P. A. (2010). Responsibilities in international research: A new look revisited. *Journal of Medical Ethics, 36*, 194–197.

Benayas, J. M. R., Newton, A. C., Diaz, A., & Bullock, J. M. (2009). Enhancement of biodiversity and ecosystem services by ecological restoration: A meta-analysis. *Science, 325*, 1121–1124.

Bergmann, S. (2009). Climate change changes religion: Space, spirit, ritual, technology—through a theological lens. *Studia Theologica, 83*, 98–118.

Bertram, C. (2009). Exploitation and intergenerational justice. In A. Gosseries & L. Meyer (Eds.), *Intergenerational Justice* (pp. 147–166). New York, NY: Oxford University Press.

Bipartisan Policy Center, Task Force on Climate Remediation Research. (2011). *Geoengineering: A national strategic plan for research on the potential effectiveness, feasibility, and consequences of climate remediation technologies*. Retrieved from http://www.bipartisanpolicy.org/sites/default/files/
BPC%20Climate%20Remediation%20Final%20Report.pdf

Bittman, M. (2011a, March 8). Sustainable farming can feed the world? *New York Times*. Retrieved from http://opinionator.blogs.nytimes.com/2011/03/08/sustainable-farming/?scp=1&sq=Bittman%20Sustainable%20March%208%202011&st=cse

Bittman, M. (2011b, September 25). Is junk food really cheaper? *New York Times*, Sunday Review, pp. 1 & 7.

Blackstock, J. J., Battisti, D. S., Caldeira, K., Eardley, D. M., Keith, D. W., Patrinos, A. A. N., ... Koonin, S. E. (2009). Climate engineering responses to climate emergencies. Retrieved from http://arxiv.org/pdf/0907.5140

Blackstock, J. & Long, L. (2010). The politics of geoengineering. *Science, 327*(5965), 527.

Bleir, R. (1984). *Science and gender: A critique of biology and its theories on women*. Elmsford, NY: Pergamon Press.

Bodansky, D. (2011, November). *Governing climate engineering: Scenarios for analysis*. Discussion Paper presented at the Harvard Project on Climate Agreements, Harvard University. Retrieved from http://papers.ssrn.com/sol3/papers.cfm?abstract_id=1963397

Bogue, R. (1989). *Deleuze and Guattari*. London, England: Routledge.

Bok, D. (2010). *The politics of happiness*. Princeton, NJ: Princeton University Press.

Bond, P. (2010). Climate debt owed to Africa: What to demand and how to collect? *Links International Journal of Socialist Renewal*. Retrieved from http://links.org.au/node/1675

Borgmann, A. (1984). *Technology and the character of contemporary life: A philosophical inquiry*. Chicago, IL: University of Chicago Press.
Borgmann, A. (2006). *Real American ethics: Taking responsibility for our country*. Chicago, IL: University of Chicago Press.
Borgmann, A. (2011). Science, ethics, and technology and the challenge of global warming. In D. Scott and B. Francis (Eds.), *Debating Science: Deliberation, Values, and the Common Good* (pp. 169–177). Amherst, NY: Humanity Books.
Bosso, C. J. (2005). *Environment inc.: From grassroots to beltway*. Lawrence, KS: University Press of Kansas.
Bracmort, K., Lattanzio, R., Barbour, E. (2010). *Geoengineering: Governance and technology policy* (Congressional Research Service Report R41371). Retrieved from http://www.fas.org/sgp/crs/misc/R41371.pdf
Brainard, L., Jones, A., & Purvis, N. (Eds.). (2009). *Climate Change and Global Poverty: A Billion Lives in the Balance?* Washington, D.C.: Brookings Institution Press.
Broad, W. J. (2006, June 27). How to cool the planet (maybe). *New York Times*. Retrieved from www.nytimes.com/2006/06/27/science/earth/27cool.html
Brooks, D. (2009, April 7). The end of philosophy. *New York Times*. Retrieved from http://nytimes.com/2009/04/07/opinion/07Brooks.html
Brovkin, V., Petoukhov, V., Claussen, M., Bauer, E., Archer, D., & Jaeger, C. (2009). Geoengineering climate by stratospheric sulfur injections: Earth system vulnerability to technological failure. *Climatic Change, 92*, 243–259.
Buber, M. (1970). *I and thou*. New York, NY: Scribners.
Buck, H. J. (2011). Climate engineering: Re-making climate for profit, or humanitarian intervention? *Development and Change, 43*, 1–18.
Bunzl, M. (2008). An ethical assessment of geoengineering. *Bulletin of the Atomic Scientists, 64*(2), 18.
Bunzl, M. (2009). Researching geoengineering: Should not or could not? *Environmental Research Letters, 4*, 1–3.
Burns, W. C. (2011). Climate geoengineering: Solar radiation management and its implications for intergenerational equity. *Stanford Journal of Law, Science & Policy, May*, 38–55.
Burrows, M. T., Schoeman, D. S., Buckley, L. B., Moore, P., Poloczanska, E. S., Brander, K. M., Richardson, A. J. (2011). The pace of shifting climate in marine and terrestrial ecosystems. *Science, 334*(6056), 652–655.
Cafaro, P. (2001a). The naturalist's virtues. *Philosophy in the Contemporary World, 8*, 85–99.
Cafaro, P. (2001b). Thoreau, Leopold, and Carson: Toward an environmental virtue ethics. *Environmental Ethics, 23*, 1–17.
Caldeira, K. (2007, October 24). How to cool the globe. *New York Times*. Retrieved from www.nytimes.com/2007/10/24/opinion/24caldiera.html
Camacho, A. E., Doremus, H., McLachlan J. S., & Minteer, B. A. (2010). Reassessing conservation goals in a changing climate. *Issues in Science Technology, 26*, 21–26.
Campaign for Tobacco–Free Kids. (2012). State tobacco taxes: A win-win-win solution. Retrieved from http://www.tobaccofreekids.org/what_we_do/state_local/taxes/
Caney, S. (2008). Human rights, climate change, and discounting. *Environmental Politics, 17*(4), 536–555.
Caney, S. (2010). Climate change, human rights, and moral thresholds. In S. Gardner, S. Caney, D. Jamieson, & H. Shue (Eds.), *Climate Ethics* (pp. 163–177). Oxford, England: Oxford University Press.
Carr, D. (2011, April 11). Keep your thumbs still when I'm talking to you. *New York Times*. Retrieved from http://www.nytimes.com/2011/04/17/fashion/17TEXT.html?scp=1&sq=%22Keep%20your%20thumbs%20still%22&st=cse.
Carson, R. (1956). *The sense of wonder*. New York, NY: Harper and Row.
Castoriadis, C. (1984). *Crossroads in the labyrinth*. Brighton, England: Harvester Press.
Castoriadis, C. (1987). *The imaginary institution of society* (K. Blamey, Trans.). Cambridge, England: Polity Press.
Chen, I. C., Hill, J. K., Ohlemüller, R., Roy, D. B.. & Thomas, C. D. (2011). Rapid range shifts of species associated with high levels of climate warming. *Science, 333*(6045), 1024–1026.

Chu, S. (2009). Carbon capture and sequestration. *Science, 325*(5948), 1599.
Cicerone, R. J. (2006). Geoengineering: Encouraging research and overseeing implementation. *Climatic Change, 77*, 221–226.
Climate Interactive. (2011). Possibilities for the global climate deal. Retrieved from http://climateinteractive.org/scoreboard/scoreboard-science-and-data/graphs-possibilities-for-the-global-climate-deal
Code, L. (1993). Taking subjectivity into account. In L. Alcoff & E. Potter (Eds.), *Feminist Epistemologies* (pp. 15–48). New York, NY: Routledge.
Cohen, R. (2012, January 2). A time to tune out. *New York Times*. Retrieved from http://www.nytimes.com/2012/01/03/opinion/cohen-a-time-to-tune-out.html?scp=1&sq=%22a%20time%20to%20tune%20out%22&st=cse
Colchester, M. & MacKay, F. (2004, August). *In search of middle ground: Indigenous peoples, collective representation and the right to free, prior and informed consent*. Paper presented at the 10th Conference of the International Association for the Study of Common Property, Oaxaca, Mexico.
Comstock, G. (2010). Ethics and genetically modified food. In F. Gottwald (Ed.), *Food Ethics* (pp. 49–66). Dordrecht, Netherlands: Springer.
Cone, J. (Producer) (2005). *Celilo Falls and the remaking of the Columbia River* [DVD]. Oregon State University.
Corner, A. & Pidgeon, N. (2010). Geoengineering the climate: The social and ethical implications. *Environment, 52*(1), 24–37.
Crutzen, P. (2006). Albedo enhancement by stratospheric sulfur injections: A contribution to resolve a policy dilemma? *Climatic Change, 77*(3–4), 211–220.
Cuomo, C. (2011). Climate change, vulnerability, and responsibility. *Hypatia, 26*, 690–714.
Dancy, J. (2010). Prima facie reasons. In J. Dancy, E. Sosa, & M. Steup (Eds.), *A Companion to Epistemology, Vol. 4* (2nd ed.) (p. 609). New York, NY: John Wiley & Sons.
David, G. V., Morgan, G., Apt, J., Steinbruner, J., & Ricke, K. (2009, March/April). The geoengineering option: A last resort against global warming? *Foreign Affairs*. Retrieved from http://fullaccess.foreignaffairs.org/20090301faessay88206/david-g-victor-granger-moragan-jay-apt-john-steinbruner-and-keat/the-geoengineering-option
Davies, G. (2006). The sacred and the profane: Biotechnology, rationality, and public debate. *Environment and Planning A, 38*(3), 423–443.
De Landa, M. (2002). *Intensive science and virtual philosophy*. London, England: Continuum.
De Landa, M. (2005). Space: Extensive and intensive, actual and virtual. In I. Buchanan & G. Lambert (Eds.), *Deleuze and Space* (pp. 80–88). Edinburgh, Scotland: University of Edinburgh Press.
Dean, C. (2011, October 4). Group urges research into aggressive efforts to fight climate change. *New York Times*, p. A18.
Deleuze, G. (1994). *Difference and repetition* (P. Patton, Trans.). London, England: Athlone Press.
Deleuze, G. & Guattari, F. (1988). *A thousand plateaus: Capitalism and schizophrenia* (B. Massumi, Trans.). London, England: Athlone Press.
Dembe, A. E. & Boden, L. I. (2000). Moral hazard: A question of morality? *New Solutions, 10*(3), 257–279.
Donner, S. D. (2007). Domain of the gods: An editorial essay. *Climatic Change, 85*, 231–236.
Dörries, M. (2010). Climate catastrophes and fear. *Wiley Interdisciplinary Reviews: Climate Change, 1*(6), 885–890.
Dowd, B. E. (1982). The logic of moral hazard: A game theoretic illustration. *The Journal of Risk and Insurance, 49*(3), 443–447.
Drengson, A. R. (1984). The sacred and the limits of the technological fix. *Zygon, 19*(3), 259–275.
Durkheim, E. (1933). *The division of labor in society*. New York, NY: Free Press.
Dyson, F. J. (1977). Can we control the carbon dioxide in the atmosphere? *Energy, 2*, 287–291.
Easterbrook, G. (2003). *The progress paradox: How life gets better while people feel worse*. New York, NY: Random House.
Edwards, D. (2006). *Ecology at the heart of faith*. Maryknoll, NY: Orbis.

Elliot, R. (1992). Intrinsic value, environmental obligation, and naturalness. *The Monist, 75*, 138–160.
Engineering and Physical Sciences Research Council (EPSRC) and Natural Environment Research Council (NERC). (2009). *Geoengineering Scoping Workshop—Outputs.* Retrieved from http://www.epsrc.ac.uk/pages/searchresults.aspx?query=geoengineering
Faulkner, E. J. (1960). *Health Insurance.* New York, NY: McGraw Hill.
Fecht, S. (2011a, September). U.K. researchers to test "artificial volcano" for geoengineering the climate. *Scientific American.* Retrieved from http://www.scientificamerican.com/article.cfm?id=uk-researchers-to-test-artificial-volcano-for-geoengineering-the-climate
Fecht, S. (2011b, October). U.K. geoengineering tests delayed until spring. *Scientific American.* Retrieved from http://blogs.scientificamerican.com/observations/2011/10/07/geoengineering-tests-delayed-until-spring/
Federal Columbia River Power System. (2001). The Columbia River system inside story (2nd ed.). Retrieved from http://www.bpa.gov/power/pg/columbia_river_inside_story.pdf
Fehr, E. & Fischbacher, U. (2004). Social norms and human cooperation. *Trends in Cognitive Sciences, 8*, 185–190.
Fischbacher, U., Gächter, S., & Fehr, E. (2001). Are people conditionally cooperative? Evidence from a public goods experiment. *Economics Letters, 71*, 397–404.
First Things. (2009). While we're at it. *First Things*, 198, 69–72.
Fleming, J. R. (2006). The pathological history of weather and climate modification: Three cycles of promise and hype. *Historical Studies in the Physical Sciences, 37*(1), 3–25.
Fleming, J. R. (2007). The climate engineers. *Wilson Quarterly, Spring*, 46–60.
Fox Keller, E. (1983). *A feeling for the organism: The life and work of Barbara McClintock.* New York, NY: Freeman.
Foden, W., Mace, G. M., Vié, J. C., Angulo, A., Butchart, S. H. M, DeVantier, L. Turak, E. (2008). Species susceptibility to climate change impacts. In J. C. Vie, C. Hilton-Taylor, & S. N. Stuart (Eds.), The 2008 Review of the International Union of Conservation of Nature Red List of Threatened Species (pp. 77–88). Gland, Switzerland: International Union for Conservation of Nature. Retrieved from http://data.iucn.org/dbtw-wpd/edocs/RL-2009-001.pdf
Frend, W. H. C. (1984). *The rise of Christianity.* Minneapolis, MN: Fortress Press.
Garcia–Perez, M., (2008). The formation of polyaromatic hydrocarbons and dioxins during pyrolysis: A review of the literature with descriptions of biomass composition, fast pyrolysis technologies and thermochemical reactions. Washington State University. Retrieved from http://pacificbiomass.org/documents/TheFormationOfPolyaromaticHydrocarbonsAndDioxinsDuringPyrolysis.pdf
Gardiner, S. M. (2001). The real tragedy of the commons. *Philosophy & Public Affairs, 30*, 387–416.
Gardiner, S. M. (2003). The pure intergenerational problem. *The Monist, 86*(3), 481–500.
Gardiner, S. M. (2004). Ethics and global climate change. *Ethics, 114*(3), 555–600.
Gardiner, S. M. (2006a). A core precautionary principle. *The Journal of Political Philosophy, 14*(1), 33–60.
Gardiner, S. M. (2006b). A perfect moral storm: Climate change, intergenerational ethics and the problem of corruption. *Environmental Values, 15*(3), 397–413.
Gardiner, S. M. (2009a). A contract on future generations? In A. Gosseries & L. Meyer (Eds.), *Intergenerational Justice* (pp. 77–118). New York, NY: Oxford University Press.
Gardiner, S. M. (2009b). Saved by disaster? Abrupt climate change, political inertia, and the possibility of an intergenerational arms race. *Journal of Social Philosophy, 40*, 140–162.
Gardiner, S. M. (2010). Is "arming the future" with geoengineering really the lesser evil? Some doubts about the ethics of intentionally manipulating the climate system. In S. Gardiner, S. Caney, D. Jamieson, & H. Shue (Eds.), *Climate Ethics: Essential Readings* (284–312). New York, NY: Oxford University Press.
Gardiner, S. M. (2011a). *A perfect moral storm: The ethical tragedy of climate change.* New York, NY: Oxford University Press.
Gardiner, S. M. (2011b). Some early ethics of geoengineering the climate: A commentary on the values of the Royal Society report. *Environmental Values, 20*(2), 163–188.
Gauthier, D. (1986). *Morals by agreement.* Oxford, England: Clarendon Press.

Giles, J. (2010). Hacking the planet: Who decides? *New Scientist, 206*(2754), 6.
Gillett, N. P., Arora, V. K., Zickfeld, K., Marshall, S. J., & Merryfield, W. J. (2011). Ongoing climate change following a complete cessation of carbon dioxide. *Nature Geoscience, 4*, 83–87.
Glickman, S. W., McHutchinson, J. G., Peterson, E. D., Cairns C. B., Harrington, R. A., Califf, R. M., & Schulman, K. A. (2009). Ethical and scientific implications of the globalization of clinical research. *New England Journal of Medicine, 360*, 816–823.
Godfray, C. (2010). Experiment earth? Report on a public dialogue on geoengineering. Retrieved from http://www.nerc.ac.uk/about/consult/geoengineering-dialogue-final-report.pdf
Goes, M., Tuana, N., & Keller, K. (2011). The economics (or lack thereof) of aerosol geoengineering. *Climatic Change, 109*(3), 719–744.
Gold, R. & Campoy, A. (2009, April 13). Oil industry braces for drop in U.S. thirst for gasoline. *Wall Street Journal.* Retrieved from http://online.wsj.com/article/SB123957686061311925.html
Goodell, J. (2010). *How to cool the planet, geoengineering and the audacious quest to fix earth's climate.* Boston, MA: Houghton Mifflin Harcourt.
Gould, C. C. (2007). Transnational solidarities. *Journal of Social Philosophy, 38*, 148–164.
Gould, S. J. (1981). *The Mismeasure of Man.* New York, NY: W. W.Norton.
Gowdy, J. M. & Olsen, P. (1994). Further problems with neoclassical environmental economics. *Environmental Ethics 16*, 161–171.
Gupta, A. (2010, June). Geoengineering the Planet? *Z Magazine, 23*(6).
Hacking, I. (1975). *The emergence of probability: A philosophical study of early ideas about probability, induction, and statistical inference.* London, England: Cambridge University Press.
Hacking, I. (2003). Risk and dirt. In R. V. Ericson & A. Doyle (Eds.), *Risk and Morality.* Toronto, ON: University of Toronto Press.
Hajer, M. (2006). The living institutions of the EU: Analysing governance as performance. *Perspectives on European Politics and Society, 7*(1), 41–55.
Hale, B. (2009). What's so moral about the moral hazard? *Public Affairs Quarterly, 23*(1), 1–26.
Hale, B. (2010, October). Moral hazards and geoengineering. Paper presented at the Ethics of Geoengineering Workshop, University of Montana.
Hale, B. (2011). Getting the bad out: Remediation technologies and respect for others. In J. K. Cambell, M. O'Rourke, & M. Slater (Eds.), *The Environment: Topics in Contemporary Philosophy, Vol. 9* (pp., 223-44). Cambridge, MA: MIT Press.
Hamilton, C. (2011a, September). *Ethical anxieties about geoengineering: Moral hazard, slippery slope, and playing god.* Paper presented at Australian Academy of Science Conference: Geoengineering the Climate? A Southern Hemisphere Perspective, Canberra, Australia. Retrieved from http://www.clivehamilton.net.au/cms/media/ethical_anxieties_about_geoengineering.pdf
Hamilton, C. (2011b, December 5). The clique that is trying to frame the global geoengineering debate. *Guardian.* Retrieved from http://www.guardian.co.uk/environment/2011/dec/05/clique-geoengineering-debate
Hamilton, C. (2011c). The ethical foundation of climate engineering. Retrieved from http://www.clivehamilton.net.au/cms/media/ethical_foundations_of_climate_engineering.pdf
Hampicke, U. (2003). The capacity to solve problems as a rationale for intertemporal discounting. *International Journal of Sustainable Development, 6*, 98–116.
Hansen, M. B. N. (2009). System–environment hybrids. In B. Clarke and M. B. N. Hansen (Eds.), *Emergence and Embodiment: New Essays on Second–Order Systems Theory* (pp. 113–142). Durham, NC: Duke University Press.
Harden, B. (1997). *A river lost: The life and death of the Columbia* (1st pbk. ed.). New York, NY: W. W. Norton.
Hardin, G. (1968). The tragedy of the commons. *Science, 162*, 1243–1248.
Harrington, B. (2001, September). Money and the moral hazard. *The American Prospect, 10*(16).

Harris, P. (2011, January 22). Social networking under fresh attack as tide of cyber-scepticism sweeps US. *Guardian*. Retrieved from http://www.guardian.co.uk/media/2011/jan/22/social-networking-cyber-scepticism-twitter

Harvey, J. (2007). Moral solidarity and empathetic understanding: The moral value and scope of the relationship. *Journal of Social Philosophy, 38*, 22–37.

Hefner, P. (1993). *The human factor*. Minneapolis, MN: Fortress Press.

Hegerl, G. C. & Solomon, S. (2009). Risks of climate engineering. *Science, 325*, 955–956.

Heidegger, M. (1977). The origin of the work of art. In *Basic writings: From "Being and Time" (1927) to "The Task of Thinking" (1964)* (pp. 143–88). New York, NY: Harper and Row.

Heyd, D. (2007). Justice and solidarity: The contractarian case against global justice. *Journal of Social Philosophy, 38*, 112–130.

Heyd, D. (2009). A value or an obligation? Rawls on justice to future generations. In A. Gosseries & L. Meyer (Eds.), *Intergenerational Justice* (pp. 167–188). New York, NY: Oxford University Press.

Hill, T. (2007). Ideals of excellence and preserving natural environments. In H. LaFollette (Ed.), *Ethics in Practice* (3rd ed.). Oxford, England: Blackwell Publishers.

Hill, W. (1987). Theology. In J. A. Komonchack & M. Collins (Eds.), *The New Dictionary of Theology* (pp. 1011–1027). Collegeville, MN: Liturgical Press.

Hodgson, L. (2010). Kant on the right to freedom: A defense. *Ethics, 120*(4), 791–819.

Hoegh-Guldberg, O., Hughes, L., McIntyre, S., Lindenmayer, D. B., Parmesan, C., Possingham, H.P., & Thomas, C. D. (2008). Assisted colonization and rapid climate change. *Science, 321*(5887), 345–346.

Hoffert, M. I., Caldeira, K., Benford, G., Criswell, D. R., Green, C., Wigley, T. M. L. (2002). Advanced technology paths to global climate stability: Energy for a greenhouse planet. *Science, 298*(5595), 981–987.

Holmström, B. (1979). Moral hazard and observability. *The Bell Journal of Economics,10*(1), 74–91.

Horgan, J. (2011, December 25). Why we lie [Review of the book *The logic of deceit and self-deception in human life*, by R. Trivers]. *New York Times Book Review*, p. 17.

House of Commons, Science and Technology Committee. (2010). *The Regulation of Geoengineering* (HC 221). London, England: The Stationary Office Limited.

Hulme, M. (2008). The conquering of climate: Discourses of fear and their dissolution. *Geographical Journal, 174*, 5–16.

Hulme, M. (2009). *Why we disagree about climate change: Understanding controversy, inaction and opportunity*. New York, NY: Cambridge University Press.

Humphreys, D. (2011). Smoke and mirrors: Some reflections on the science and politics of geoengineering. *The Journal of Environment Development, 20*, 99–120.

Hunter, M. L. (2001). *Fundamentals of conservation biology*. Oxford, England: Blackwell Publishers.

Ingold, T. (2010a). Bringing things to life: Creative entanglements in a world of materials (National Centre for Research Methods Working Paper 15). Retrieved from http://eprints.ncrm.ac.uk/1306/1/0510_creative_entanglements.pdf

Ingold, T. (2010b). The textility of making. *Cambridge Journal of Economics, 34*(1), 91–102.

Integrated Assessment of Geoengineering Proposals. (2011). Integrated assessment of geoengineering proposals (IAGP) workshops for stakeholders 2011. Retrieved from www.iagp.ac.uk/iagp-workshops-stakeholders-2011

Intemann, K. (2009). Why diversity matters: Understanding and applying the diversity component of the NSF's broader impacts criterion. *Social Epistemology, 23*, 249–266.

Intergovernmental Panel on Climate Change (IPCC). (2007a). Climate change 2007: Synthesis report. Contribution of working groups I, II, and III to the fourth assessment report of the Intergovernmental Panel on Climate Change, R. K Pachauri & A. Reisinger, (Eds.). Geneva, Switzerland: Intergovernmental Panel on Climate Change.

Intergovernmental Panel on Climate Change (IPCC). (2007b). Solomon, S., D. Qin, M. Manning, Z. Chen, M. Marquis, K. B. Averyt, M. Tignor and H. L. Miller (Eds.), *Climate Change 2007: The Physical Science basis. Contribution of Working Group I to the Fourth Assessment Report of the Intergovernmental Panel on Climate Change.* Cambridge, UK and New York, USA: Cambridge University Press.

Intergovernmental Panel on Climate Change (IPCC). (2007c). M. L. Parry, O. F. Canziani, J. P. Palutikof, P. J. van der Linden, & C. E. Hanson (Eds.), *Climate change 2007: Impacts, adaptation and vulnerability. contribution of working group II to the fourth assessment report of the intergovernmental panel on climate change.* Cambridge, UK and New York, USA: Cambridge University Press.

Intergovernmental Panel on Climate Change (IPCC). (2007d). Summary for Policymakers. In: B. Metz, O. R. Davidson, P. R. Bosch, R. Dave, L. A. Meyer (Eds.), *Climate change 2007: Mitigation. contribution of working group III to the fourth assessment report of the intergovernmental panel on climate change* Cambridge, UK and New York, USA: Cambridge University Press.

Intergovernmental Panel on Climate Change (IPCC). (2010, October). *The IPCC fifth assessment report (AR5): Proposal for an IPCC expert meeting on geoengineering* (IPCC-XXXII/Doc. 5). Busan, South Korea.

International Labour Organization. (1989). Convention on Indigenous and Tribal Peoples (C169). Retrieved from http://www.ilo.org/ilolex/cgi-lex/convde.pl?C169

Irvine, P., Ridgwell, A., & Lunt, D. (2011). Climate effects of surface albedo geoengineering. *Journal of Geophysical Research, 116,* D24112.

Israel, J. I. (2011). *Democratic enlightenment: Philosophy, revolution, and human rights 1750–1790.* New York, NY: Oxford University Press.

Iyer, P. (2012, January 1). The joy of quiet. *New York Times,* Sunday Review, pp. 1, 6.

James, S. D. (2009, February 18). Americans hunker down in recession. *ABC News.* Retrieved from http://abcnews.go.com/Business/Economy/story?id=6858500&page=1#.T0g-EfGqD_M

Jamieson, D. (1992). Ethics, public policy, and global warming. *Science, Technology,and Human Values, 17*(2), 139–153.

Jamieson, D. (1996). Ethics and intentional climate change. *Climatic Change, 33*(3), 323–336.

Jamieson, D. (2005). Adaptation, mitigation, and justice. In W. Sinnott–Armstrong & R. B. Howarth (Eds.), *Perspectives on Climate Change: Science, Economics, Politics, Ethics, Vol. 5: Advances in the Economics of Environmental Resources* (pp. 217–248). London, England: Elsevier Ltd.

Jamieson, D. (2010, September). Can space reflectors save us? Why we shouldn't buy into geoengineering fantasies. *Slate.* Retrieved from http://www.slate.com/articles/technology/future_tense/2010/09/can_space_reflectors_save_us.html

Jänicke, M. (2008). *Megatrend umweltinnovation: zur ökologischen modernisierung von wirtschaft und staat.* Munich, Germany: Oekom.

Jänicke, M. (2012). Dynamic governance of clean-energy markets: How technical innovation could accelerate climate policies. *Journal of Cleaner Production, 22,* 50–59.

Jarvis, A., Leedal, D., Taylor, J., & Young, P. (2009). Stabilizing global mean surface temperature: A feedback control perspective. *Environmental Modelling & Software, 24*(5), 665–74.

Johnson, B. (2003). Ethical obligations in a tragedy of the commons. *Environmental Values, 12,* 271–287.

Juster, F. T. & Stafford F. P. (1985). *Time, goods, and well-being.* Ann Arbor, MI: Institute for Social Research, The University of Michigan.

Kahneman, D. & Tversky, A. (1984). Choices, values, and frames. *American Psychologist, 39,* 341–50.

Karl, T. R., Meehl, G. A., Peterson, T. C., Kunkel, K. E., Gutowski Jr., W. J., & Easterling, D. R. (2008). Executive summary. In T. R. Karl, G. A. Meehl, C. D. Miller, S. J. Hassol, A. M. Waple, & W. L. Murray (Eds.), *Weather and Climate Extremes in a Changing Climate. Regions of Focus: North America, Hawaii, Caribbean, and U.S. Pacific Islands.* Washington, D.C.: U.S. Climate Change Science Program and the Subcommittee on Global Change Research.

Katz, E. (1992). The call of the wild: The struggle for domination and the technological fix of nature. *Environmental Ethics, 14*(3), 265–273.
Katz, E. (2000). The big lie. In W. Throop (Ed.), *Environmental Restoration* (pp. 83–93). Amherst, NY: Humanity Books.
Keesing, F., Belden, L. K., Daszak, P., Dobson, A., Harvell, C. D., Holt., R. D., Ostfeld, R. S. (2010). Impacts of biodiversity on the emergence and transmission of infectious diseases. *Nature, 468*, 647–652.
Keith, D. W. (2000). Geoengineering the climate: History and prospect. *Annual Review of Energy and the Environment, 25,* 245–84.
Keith, D. W. (2010). Engineering the planet. In S. Schneider, A. Rosencrantz, M. Mastrandrea, & K. Kuntz–Duriseti (Eds.), *Climate change science and policy*. Washington, D.C.: Island Press.
Keith, D. W., Parson, E., & Morgan, M. G. (2010). Research on global sun block needed now. *Nature, 463,* 426–427.
Kellogg, W. & Schneider. S. (1974). Climate stabilization: For better or for worse? *Science, 186,* 1163–1172.
Kempton, W., Boster, J. S., & Hartley, J. A. (1995). *Environmental values in American culture*. Cambridge, MA: MIT Press.
Kiehl, J. (2006). Geoengineering climate change: Treating the symptom over the cause? *Climatic Change, 77*(3), 227–228.
Kumar, R. (2009). Wronging future people: A contractualist proposal. In A. Gosseries & L. Meyer (Eds.), *Intergenerational Justice* (pp. 251–272). New York, NY: Oxford University Press.
Lane, L. (2009, November 5). Researching solar radiation management as a climate policy option: Statement before the house committee on science and technology, 111th Congress.
Lane, R. E. (2000). *The loss of happiness in market democracies.* New Haven, CT: Yale University Press.
Latour, B. (2002). Morality and technology: The end of the means. *Theory, Culture, and Society, 19*(5–6), 247–60.
Leach, M., Fairhead, J., & Fraser, J. (2011, April). Land grabs for biochar? Narratives and counter biogenic carbon sequestration economy. Paper presented at the International Conference on Global Land Grabbing, Sussex, England.
Leach, M., Scoones, I., & Wynne, B. (2005). *Science and citizens: Globalization and the challenge of engagement*. London, England: Zed Books.
LeCain, T. L. (2004). When everybody wins does the environment lose? The environmental techno-fix in twentieth-century mining. In L. Rosner (Ed.), *The Technological Fix: How people use technology to create and solve social problems* (pp. 137–153). New York, NY: Routledge.
Lee, K. (1999). *The natural and the artefactual: The implications of deep science and deep technology for environmental philosophy*. Lanham, MD: Lexington Books.
Leiserowitz, A. (2007/2008). International public opinion, perception, and understanding of global climate change (Human Development Report Office Occasional Paper). In Human development report: Fighting climate change: Human Solidarity in a Divided World. Retrieved from http://hdr.undp.org/en/reports/global/hdr2007-8/papers/leiserowitz_anthony6.pdf
Leiserowitz, A. (2010, March). Geoengineering and climate change in the public mind. Paper presented at Asilomar International Conference on Climate Intervention Technologies, Pacific Grove, California.
Lempert, R. J. & Prosnitz, D. (2011). *Governing geoengineering research: A political and technical vulnerability analysis of potential near–term options* (RAND Corporation document TR846). Retrieved from http://www.rand.org/pubs/technical_reports/TR846
Lempert, R. J. & Schlesinger, M. E. (2001). Robust strategies for abating climate change (RAND Corporation document RP904). Retrieved from http://www.rand.org/pubs/reprints/RP904.html

Levene, M. (2010). The apocalyptic as contemporary dialectic: From Thanatos (violence) to Eros (transformation). In S. Skrimshire (Ed.), *Future Ethics: Climate Change and Apocalyptic Imagination* (pp. 59–80). New York, NY: Continuum.

Lohmann, L. (2006) Carbon trading: A critical conversation on climate change, privitisation and power. *Development Dialogue, 48*. Retrieved from http://www.dhf.uu.se/pdffiler/DD2006_48_carbon_trading/carbon_trading_web.pdf

Long, J. (2010 March 2010). *Statement to the house science committee, geoengineering III hearings, 112th Congress*. Retrieved from http://archives.democrats.science.house.gov/Media/file/Commdocs/hearings/2010/Full/18mar/Long_Testimony.pdf

Longino, H. (2002). *The fate of knowledge*. Princeton, NJ: Princeton University Press.

Lovelock, J. (2008). A geophysiologist's thoughts on geoengineering. *Philosophical Transactions of the Royal Society A, 366*, 3883–3890.

Lovett, F. (2010). *A general theory of domination and justice*. New York, NY: Oxford University Press.

Lunt, D. J., Ridgwell, A., Valdes, P. J., & Seale, A. (2008). "Sunshade world": A fully coupled GCM evaluation of the climatic impacts of geoengineering. *Geophysical Research Letters, 35*(12), 2–6.

Lynas, M. (2011). *The god species: How the planet can survive the age of humans*. London, England: Fourth Estate.

MacKay, F. (2004). Indigenous peoples' right to free, prior and informed consent and the world bank's extractive industries review. *Sustainable Development Law & Policy, IV*(2), 43–65.

Mackenzie, A. (2002). *Transductions: Bodies and machines at speed*. New York, NY: Continuum.

MacNaghten, P. & Owen, R. (2011). Environmental science: Good governance for geoengineering. *Nature, 479*(7373), 293.

Maestre, F. T., Quero, J. T., Gotelli, N. J., Escudero, A., Ochoa, V., Delgado-Baquerizo, M., Zaady, E. (2012). Plant species richness and ecosystem multifunctionality in global drylands. *Science, 335*(6065), 214–218.

Magurran, A. E. and Dornelas, M. (2010). Biological diversity in a changing world. *Philosophical Transactions of the Royal Society B, 365*, 3593–3597.

Marland, G. (1996). Could we/should we engineer the earth's climate? *Climatic Change, 33*(3), 275–278.

Marshall, J. M. (1976). Moral hazard. *American Economic Review, 66*(5), 880–890.

Matthews, H. D. and Caldeira, K. (2007). Transient climate—carbon simulations of planetary geoengineering. *Proceedings of the National Academy of Sciences, 104*, 9949–9954.

May, L. (1996). *The socially responsive self: Social theory and professional ethics*. Chicago, IL: University of Chicago Press.

May, L. (2007). The international community, solidarity, and the duty to aid. *Journal of Social Philosophy 38*, 185–203.

Meece, M. (2011, February 6). Who's the boss, you or your gadget? *New York Times*, pp. 1, 8.

Mercer, A. & Keith, D. W. (2010, October). Engineering scientific inquiry and solar radiation management research. Paper presented at the Ethics of Geoengineering Workshop, University of Montana. Retrieved from http://www.umt.edu/ethics/EthicsGeoengineering/Workshop/articles1/Ashely%20Mercer.pdf

Mercer, A., Keith, D. W., & Sharp, J. (2011). Public understanding of solar radiation management. *Environmental Research Letters, 6*, 1–9.

Metcalf, C. (2004). Indigenous rights and the environment: Evolving international law. *Ottawa Law Review, 35*(1), 101–140.

McFague, S. (2008). *A new climate for theology*. Minneapolis, MN: Fortress Press.

McGuckin, J. A. (2004). *The Westminster handbook to patristic theology*. Louisville, KY: Westminster John Knox Press.

McKibben, B. (1989). *The end of nature*. New York, NY: Random House.

McKinsey & Company & the Conference Board. (2007, December). Reducing U.S. greenhouse gas emissions: How much at what cost? Retrieved from http://www.mckinsey.com/Client_Service/Sustainability/Latest_thinking/Reducing_US_greenhouse_gas_emissions

McLaren, D. (2011). Negatonnes—An initial assessment of the potential for negative emission techniques to contribute safely and fairly to meeting carbon budgets in the 21st century. Retrieved from http://www.foe.co.uk/resource/reports/negatonnes.pdf

McMahon, D. (2011, December 25). The Enlightenment's true radicals. [Review of the book *Democratic enlightenment: Philosophy, revolution, and human rights 1750–1790*, by J. I. Israel]. *New York Times Book Review*, p. 12.

McPhee, J. (1989). *The control of nature*. New York, NY: Farrar, Strauss and Giroux.

Mill, J. S. (1904). On nature. In *Nature: The utility of religion, and theism* (pp. 7–33). London, England: Watts and Co.

Minteer, B. and Collins, J. (2010). Move it or lose it? The ecological ethics of relocating species under climate change. *Ecological Applications, 20*, 1801–1804.

Moreno-Cruz, J., Ricke, K., & Keith, D. W. (2012). A simple model to account for regional inequalities in the effectiveness of solar radiation management. *Climatic Change, 110*(3), 649–668.

Morgan, M. G. & Ricke, K. (2010). Cooling the earth through solar radiation management: The need for research and an approach to its governance. Retrieved from http://www.irgc.org/IMG/pdf/SRM_Opinion_Piece_web.pdf

Moser, S. C. & Dilling, L. (2004). Making climate hot: Communicating the urgency and challenge of global climate change. *Environment, 46*, 32–46.

Myers, D. G. (2000). *The American paradox: Spiritual hunger in an age of plenty*. New Haven, CT: Yale University Press.

Nagel, T. (1993). Moral Luck. In D. Statman (Ed.), *Moral Luck* (pp 57–71). Albany, NY: State University of New York Press.

Nadasdy, P. (2005). The anti-politics of TEK: The institutionalization of co-management discourse and practice. *Anthropologica, 47*(2), 215–231.

Newport, F. (June 3, 2011). More than nine in ten Americans continue to believe in God. Retrieved from http://www.gallup.com/poll/147887/americans-continue-believe-god.aspx

Niebuhr, R. (1964). *The nature and destiny of man, Vol 1*. New York, NY: Scribners.

Niemeyer, U., Schmidt, H., & Timmreck, C. (2010). The dependency of geoengineered sulfate aerosol on the emission strategy. *Atmospheric Science Letters, 12*, 189–194.

Nisbet, M. C. (2009a). Communicating climate change: Why frames matter. *Environment, 51*(2), 12–23.

Nisbet, M. C. (2009b). Framing science: A new paradigm in public engagement. In L. Kahlor & P. Stout (Eds.), *Communicating Science: New Agendas in Communication* (pp. 40–67). New York, NY: Routledge.

Nisbet, M. C. & Scheufele, D. A. (2009). What's next for science communication? *American Journal of Botany, 96*(10), 1767–1778.

Nolt, J. (2011a). Greenhouse gas emission and the domination of posterity. In D. Arnold (Ed.), *The Ethics of Global Climate Change* (pp. 60–76). New York, NY: Cambridge University Press.

Nolt, J. (2011b). Nonanthropocentric climate ethics. *Wiley Interdisciplinary Reviews: Climate Change, 2*(5), 701–711.

Northcott, M. (2007). *A moral climate: The ethics of global warming*. Maryknoll, NY: Orbis.

Northwest Power and Conservation Council. (2010). Floods and flood control. Retrieved from http://www.nwcouncil.org/history/floods.asp

Olson, R. L. (2011). Geoengineering for decision makers. Retrieved from http://www.wilsoncenter.org/sites/default/files/Geoengineering_for_Decision_Makers_0.pdf

O'Neill, J. (1993). *Ecology, policy, and politics: Human well-being and the natural world*. London, England: Routledge.

Ostrom, E. (2000). Collective action and the evolution of social norms. *The Journal of Economic Perspectives, 14*, 137–158.

Ostrom, E. (2010). Analyzing collective action. *Agricultural Economics, 41*(s1), 155–166.

Ott, K. (2003). Reflections on discounting: Some philosophical remarks. *International Journal of Sustainable Development, 6*, 7–24.

Ott, K. (2004). Essential components of future ethics. In R. Döring & M. Rühs (Eds.), *Ökonomische Rationalität und praktische Vernunft* (pp. 83–108). Würzburg, Germany: Königshausen & Neumann.

Ott, K. (2008). Ethical foundations of climate change policies. In S. Bergmann & D. Gerthen (Eds.), *Religion and Dangerous Environmental Change* (pp. 195–204). Münster, Germany: LIT Publishers.

Ott, K., Baatz, C., & Berg, M. (2012). *Risikobewertung, bewertungsansätze und entscheidungskriterien.* Büro für Technikfolgenabschätzung beim deutschen Bundestag (Report on behalf of the Office of Technological Assessment of the German Parliament). Berlin, Germany.

Otto, R. (1958). *The idea of the holy.* New York, NY: Oxford University Press.

Palmer, C. (2011). Does nature matter? The place of the non–human in the ethics of climate change. In D. G. Arnold (Ed.), *The Ethics of Global Climate Change* (pp. 272–291). Cambridge, England: Cambridge University Press.

Paludneviciene, R. & Leigh, I. W. (Eds.). (2011). *Cochlear implants: Evolving perspectives.* Washington, DC: Gallaudet University Press.

Parfit, D. (1984). *Reasons and persons.* Oxford, England: Oxford University Press.

Parkhill, K. & Pidgeon, N. (2011). *Public engagement on geoengineering research: Preliminary report on the SPICE deliberative workshops* (Understanding Risk Working Paper 11–01, Cardiff University). Retrieved from: http://psych.cf.ac.uk/understandingrisk/docs/spice.pdf

Parthasarathy, S., Rayburn, L., Anderson, M., Mannisto, J., Maguire, M., & Najib, D. (2010). Geoengineering in the Arctic: Defining the governance dilemma (Science, Technology, and Public Policy Program Working Paper 10–3, Gerald R. Ford School of Public Policy, University of Michigan).

Pauly, M. V. (1968). The economics of moral hazard: Comment. *American Economic Review, 58*(3), 531–537.

Peterson, A. L. (2001). *Being human: Ethics, environment, and our place in the world.* Berkeley, CA: University of California Press.

Pettit, P. (1997). *Republicanism: A theory of freedom and government.* New York, NY: Oxford University Press.

Pongratz, J., Lobell, D. B., Cao, L., & Caldeira, K. (2012). Crop yields in a geoengineered climate. *Nature Climate Change, 2*(2), 1–5. Retrieved from http://www.nature.com/nclimate/journal/vaop/ncurrent/pdf/nclimate1373.pdf

Posner, E. A. & Weisbach, D. (2010). *Climate change justice.* Princeton, NJ: Princeton University Press.

Preston, C. J. (2008). Synthetic biology: Drawing a line in Darwin's sand. *Environmental Values, 17,* 23–39.

Preston, C.J. (2011). Re–thinking the unthinkable: Environmental ethics and the presumptive argument against geoengineering. *Environmental Values, 20*(4), 457–479.

Prigogine, I. & Glandsorff, P. (1971). *Thermodynamic theory of structure, stability and fluctuations.* New York, NY: Wiley.

Protevi, J. (2001). *Political physics: Deleuze, Derrida, and the body politic.* London, England: Athlone Press.

Pugliese, A. & Ray, J. (2009). A heated debate: Global attitudes toward climate change. *Harvard International Review, 31,* 64–8.

Putnam, R. D. (2000). *Bowling alone.* New York, NY: Simon & Schuster.

Rasmussen, E. (2001). *Games and information.* Malden, MA: Blackwell Publishers.

Reardon, J. (2007). Democratic mis–haps: The problem of democratization in a time of biopolitics. *BioSocieties, 2*(2), 239–256.

Rehg, W. (2007). Solidarity and the common good: An analytic framework. *Journal of Social Philosophy 38,* 7–21.

Reinhardt, B. H. (2011). Drowned towns in the cold war west: Small communities and federal water projects. *The Western Historical Quarterly, 42*(2), 149–172.

Reiss, M. J. & Straughan, R. (2001). *Improving nature? The science and ethics of genetic engineering* (Canto ed.). Cambridge, England: Cambridge University Press.

Rendall, M. (2011). Non–identity, sufficiency, and exploitation. *Journal of Political Philosophy, 19*(2), 229–247.
Reynolds, L. & Szerszynski, B. (2007). The role of participation in a techno–scientific controversy (Participatory Governance and Institutional Innovation Contract No. CIT2-CT-2004-505791, Deliverable No. 16: 6th EU Framework Program for Research and Technology). Retrieved from: http://csec.lancs.ac.uk/docs/PAGANINI_WP6_Final_Report.pdf
Richardson, D. M., Hellmann, J. J., McLachlan, J. S., Sax, D. F., Schwartz, M. W., Gonzalez, P., Vellend, M. (2009). Multidimensional evaluation of managed relocation. *Proceedings of the National Academy of Science, 106*(24), 9721–9724.
Ricke, K., Morgan, G., & Allen, M. (2010). Regional climate response to solar–radiation management. *Nature Geoscience, 3*(8), 537–541.
Ridgwell, A., Singarayer, J., Hetheringon, A., & Valdes, P. (2009). Tackling regional climate change by leaf albedo bio–geoengineering. *Current Biology 19*(2), 146–150.
Robb, C. (2010). *Wind, sun, soil, spirit: Biblical ethics and climate change.* Minneapolis, MN: Fortress Press.
Robock, A. (2008). 20 reasons why geoengineering may be a bad idea. *Bulletin of the Atomic Scientists, 64*(2), 14–8.
Robock, A., Oman, L., & Stenchikov, G. L. (2008). Regional climate responses to geoengineering with tropical and Arctic SO_2 injections. *Journal of Geophysical Research, 113*, D16101.
Robock, A., Marguardt, A., Kravitz, B., & Stenchikov, G. L. (2009). Benefits, risks, and costs of stratospheric geoengineering. *Geophysical Research Letters, 36*, L19703.
Robock, A, Bunzl, M., Kravitz, B., & Stenchikov, G. L. (2010). Atmospheric science: A test for geoengineering? *Science, 327*, 530–531.
Rogers–Hayden, T. & Pidgeon, N. (2007). Moving engagement "upstream"? Nanotechnologies and the Royal Society and Royal Academy of Engineering's inquiry. *Public Understanding of Science, 16*(3), 345–364.
Rolston III, H. (1982). Are values in nature subjective or objective? *Environmental Ethics, 4*, 125–151.
Rolston III, H. (1989). *Environmental ethics, duties to and values in the natural world.* Philadelphia, PA: Temple University Press.
Rolston III, H. (1995). Duties to endangered species. *Encyclopedia of Environmental Biology, 1*, 517–528.
Rolston III, H. (2001). Biodiversity. In D. Jamieson (Ed.), *A Companion to Environmental Philosophy* (pp. 402–415). Oxford, England: Blackwell Publishers.
Rosner, L. (2004). Introduction. In L. Rosner (Ed.), *The Technological Fix: How People Use Technology to Create and Solve Problems* (pp. 1–9). New York, NY: Routledge.
Ross, A. & Matthews, H. D. (2009). Climate engineering and the risk of rapid climate change. *Environmental Research Letters, 4*(4), 1–6.
Ross, W. D. (1930). *The right and the good.* Oxford, England: Clarendon Press.
Royal Society. (2009). Geoengineering the climate: Science, governance, and uncertainty (Royal Society Policy document 10/09). Retrieved from http://royalsociety.org/policy/publications/2009/geoengineering–climate/
Russell, L., Rasch, P., Mace, G., Jackson, R., Shepherd, J., Liss, P., . . . Morgan, M. (in press). Ecosystem Impacts of Geoengineering: A Review for Developing a Science Plan. *Ambio.*
Safire, W. (2003). *No uncertain terms.* New York, NY: Simon & Schuster.
Sandler, R. (2007). *Character and environment: A virtue–oriented approach to environmental ethics.* New York, NY: Columbia University Press.
Sandler, R. (2010). The value of species and the ethical foundations of assisted colonization. *Conservation Biology, 24*, 424–31.
Sandler, R. (in press a). The ethics of species. Cambridge, England: Cambridge University Press.
Sandler, R. (in press b). The ethics of mitigation. M. Di Paola, & G. Pellegino (Eds.). The ethics and politics of climate change. London, England: Routledge.
Santmire, H. P. (1985). *The travail of nature: The ambiguous ecological promise of Christianity.* Minneapolis, MN: Fortress Press.

Sarewitz, D. (2004). How science makes environmental controversies worse. *Environmental Science and Policy, 7,* 385–403.
Sarkar, S. (2005). *Biodiversity and environmental philosophy: An introduction.* Cambridge, England: Cambridge University Press.
Schaefer, J. (2009). *Theological foundations for environmental ethics.* Washington, DC: Georgetown University Press.
Schelling, T. C. (1996). The economic diplomacy of geoengineering. *Climatic Change, 33,* 303–307.
Schellnhuber, J. (2011). Geoengineering: The good, the MAD, and the sensible. *Proceedings of the National Academy of Sciences, 108*(51), 20428–20433.
Schleiermacher, F. (1989). *The Christian Faith.* Edinburgh, Scotland: T&T Clark Ltd.
Schleiermacher, F. (1996). *On religion: Speeches to its cultured despisers.* Cambridge, England: Cambridge University Press.
Schneider, S. H. (2001). Earth systems engineering and management. *Nature, 409,* 417–21.
Schneider, S. H. (2008). Geoengineering: Could we or should we make it work? *Philosophical Transactions of the Royal Society A, 366,* 3843–62.
Schwartz, J. (2007). From domestic to global solidarity: The dialectic of the particular and universal in the building of social solidarity. *Journal of Social Philosophy, 38,* 131–147.
Scott, D., (2011). The technological fix criticisms and the agricultural biotechnology debate. *Agriculture and Environmental Ethics, 24*(3), 207–226.
Searle, J. R. (1978). *Prima facie* obligations. In J. Raz (Ed.), *Practical Reasoning* (pp. 81–90). Oxford, England: Oxford University Press.
Seitz, R. (2010). Bright Water-hydrosols, water conservation and climate change. *Climatic Change, 105*(3–4), 15.
Seligman, M. E. P. (2002). *Authentic happiness.* New York, NY: Free Press.
Shackley, S., Young, P., Parkinson, S., & Wynne, B. (1998). Uncertainty, complexity and concepts of good science in climate change modeling: Are GCMs the best tools? *Climatic Change, 38*(2), 159–205.
Shao, Y., Peng, G., & Leslie, L. M. (2002). The environmental dynamic system. In G. Peng, L. M. Leslie, & Y. Shao (Eds.), *Environmental Modelling and Prediction* (pp. 21–74). Berlin, Germany: Springer.
Shavell, S. (1979). On moral hazard and insurance. *The Quarterly Journal of Economics, 93*(4), 541-65.
Sher, G. (1987). *Desert.* Princeton, NJ: Princeton University Press.
Shue, H. (1992). The unavoidability of justice. In A. Hurrell & B. Kingsbury (Eds.), *The International Politics of the Environment* (pp. 373–397). Oxford, England: Oxford University Press.
Shue, H. (2010). Global environment and international inequality. In S. Gardiner, S. Caney, D. Jamieson, & H. Shue (Eds.), *Climate Ethics: Essential Readings* (pp. 101–111). New York, NY: Oxford University Press.
Simondon, G. (1964). *L'individu et sa genèse physico–biologique.* Paris, France: Presses Universitaires de France.
Simondon, G. (1989). *Du mode d'existence des objets techniques* (3rd ed.). Paris, France:Aubier.
Sinervo, B., Méndez–de–la–Cruz, F., Miles, D. B., Heulin, B., Bastiaans, E., Villagrán–Santa Cruz, M. Sites Jr., J. W. (2010). Erosion of lizard diversity by climate change and altered thermal niches. *Science, 328*(5980), 894–899.
Singer, P. (2002). One atmosphere. In P. Singer (Ed.), *One World: The Ethics of Globalization* (pp. 14–50). New Haven, CT: Yale University Press.
Singer, P. (2009). *Animal Liberation* (Reissue ed.). New York, NY: Harper Perennial Modern Classics.
Skrimshire, S. (2010). Introduction: How should we think about the future? In S. Skrimshire (Ed.), *Future Ethics: Climate Change and Apocalyptic Imagination* (pp. 1–10). New York, NY: Continuum.

Solar Radiation Management Governance Initiative (SRMGI). (2011). Solar radiation management: The governance of research. Retrieved from http://www.srmgi.org/files/2012/01/DES2391_SRMGI-report_web_11112.pdf

Soto, R. (2009, August 4). My so–called digital–free life. *USA Today*, p. 9B.

Soulé, M. E. (1985). What is conservation biology? *Bioscience, 35*, 727–734.

Spence, A., Venables, D., Pidgeon, N., Poortinga, W., & Demski, C. (2010). Public perceptions of climate change and energy futures in Britain: Summary findings of a survey conducted. *in January–March 2010* (Understanding Risk Working Paper 10–01). Cardiff University: School of Psychology.

Stern, N. (2007). *The economics of climate change: The Stern review*. Cambridge, England: Cambridge University Press.

Stern, N. (2010, June 24). Climate: What you need to know [Review of *Earth: Making a life on a tough new planet*, by B. McKibben]. *New York Review of Books*, 35–37.

Stigler, G. (1971). The theory of economic regulation. *Bell Journal of Economics and Management Science, 2*, 3–21.

Stirling, A. (2005). Opening up or closing down? Analysis, participation and power in the social appraisal of technology. In M. Leach, I. Scoones, & B. Wynne (Eds.), *Science and Citizens: Globalization and the Challenge of Engagement* (pp. 218–231). London, England: Zed Books.

Stirling, A. (2008). "Opening up" and "closing down": Power, participation, and pluralism in the social appraisal of technology. *Science, Technology, & Human Values, 33*(2), 262–294.

Stone, D. (1999–2000). Beyond moral hazard: Insurance as moral opportunity. *Connecticut Insurance Law Journal, 6*(1), 11–46.

Surowiecki, J. (2010, October). Later: What does procrastination tell us about ourselves? *New Yorker*, pp. 110–113.

Sunstein, C. (2005). *Laws of fear: Beyond the precautionary principle*. New York, NY: Cambridge University Press.

Svoboda, T., Keller, K., Goes, M., & Tuana, N. (2011). Sulfate aerosol geoengineering: The question of justice. *Public Affairs Quarterly, 25*(3), 157–180.

Swart, R., Marinova, N., Bakker S., & van Tilburg, X. (2009). *Policy options to respond to rapid climate change*. Wageningen, The Netherlands: Alterra.

Szabo, L. (2004, May 5). Health experts recommend that good home cooking. *USA Today*. Retrieved from http://www.usatoday.com/life/lifestyle/2004–05–05–home–cooking_x.htm

Taylor, P. (1986). *Respect for nature*. Princeton, NJ: Princeton University Press.

Tenner, E. (1996). *Why things bite back: Technology and the revenge of unintended consequences*. New York, NY: Knopf.

Thernston, S. (2010, March). What role for geoengineering? The American-The Online Magazine of the American Enterprise Institute. Retrieved from http://www.american.com/archive/2010/march/what–role–for–geoengineering/?

Thomas, C. D. (2011). Translocation of species, climate change, and the end of trying to recreate past ecological communities. *Trends in Ecology and Evolution, 26*, 216–221.

Thomas, C. D., Cameron, A., Green, R. E., Bakkenes, M., Beaumont, L J., Collingham, Y. C., Williams, S. E. (2004). Extinction risk from climate change. *Nature, 427*, 145–148.

Thompson, J. (2009). Identity and obligation in a transgenerational polity. In A. Gosseries & L. Meyer (Eds.), *Intergenerational Justice* (pp. 25–49). New York, NY: Oxford University Press.

Tillich, P. (1951). *Systematic Theology, Vol. 1*. Chicago, IL: University of Chicago Press.

Trenberth, K. E. & Dai, A. (2007). Effects of Mount Pinatubo volcanic eruption on the hydrological cycle as an analog of geoengineering. *Geophysical Research Letters, 34* (L15702).

Trick, C., Bill, B. D., Cochlan, W. P., Wells, M. L., Trainer, V. L., & Pickell, L. D. (2010). Iron enrichment stimulates toxic diatom production in high–nitrate, low–chlorophyll areas. *Proceedings of the National Academy of Sciences, 107*, 20762–20767.

Trivers, R. (2011). *The folly of fools: The logic deceit and self–deception in human life*. New York, NY: Basic Books.

Turnbull, D. (2000). *Masons, tricksters and cartographers*. Amsterdam, The Netherlands: Harwood Academic.

Ulin, D. L. (2009, August 9). The lost art of reading. *Los Angeles Times*. Retrieved from http://www.latimes.com/entertainment/news/arts/la-ca-reading9-2009aug09,0,4905017.story

Ulrich, R. (2007). *Empty nets: Indians, dams, and the Columbia River* (2nd ed.). Corvallis, OR: Oregon State University Press.

United Nations Convention on Biological Diversity (UNCBD). (2010). Biodiversity and climate change (COP 10 Decision X/33). Retrieved from http://www.cbd.int/decision/cop/?id=12299

United Nations Development Program (UNDP). (2007). *Human development report 2007/ 2008: Fighting climate change: Human solidarity in a divided world.* New York, NY: Palgrave Macmillan.

United Nations Framework Convention on Climate Change (UNFCCC). (1992). United Nations framework convention on climate change. Retrieved from http://unfccc.int/resource/docs/convkp/conveng.pdf

United Nations Framework Convention on Climate Change (UNFCCC). (2007). Climate change: Impacts, vulnerabilities, and adaptations in developing countries. Retrieved from http://unfccc.int/resource/docs/publications/impacts.pdf

United Nations Economic and Social Council (UNESC) (1994). Report of the sub–commission on the prevention of discrimination and protection of minorities on its forty–sixth session. Retrieved from http://http://daccess-dds-ny.un.org/doc/UNDOC/GEN/G94/145/56/PDF/G9414556.pdf?OpenElement

Unnerstall, H. (1999). *Rechte zukünftiger generationen.* Würzburg, Germany: Königshausen & Neumann.

Urban, M. C., Tewksbury, J. J., & Sheldon, K. S. (2012). On a collision course: competition and dispersal differences create no-analog communities and cause extinctions during climate change. *Proceedings of the Royal Society B*. doi: 10.1098/rspb.2011.2367

U.S. Climate Change Science Program. (2008). Preliminary review of adaptation options for climate–sensitive ecosystems and resources. Retrieved from http://www.climatescience.gov/Library/sap/sap4-4/final-report/#finalreport

U.S. Department of Labor, Bureau of Labor Statistics. (2006). 100 Years of U.S. Consumer Spending. Retrieved from http://www.bls.gov/opub/uscs/

U.S. Department of Labor, Bureau of Labor Statistics. (2009). Consumer Expenditures in 2009. Retrieved from the Bureau of Labor Statistics website: http://www.bls.gov/cex/

U.S. Geological Survey. (1998). The cataclysmic 1991 eruption of Mount Pinatubo, Philippines (Fact Sheet 113–97). Retrieved from http://pubs.usgs.gov/fs/1997/fs113-97/

U.S. Government Accountability Office (U.S. GAO). (2010). A coordinated strategy could focus federal geoengineering research and inform governance efforts (GAO-10-903). Retrieved from http://www.gao.gov/assets/320/310105.pdf

U.S. Government Accountability Office (U.S. GAO). (2011). Climate engineering: Technical status, future directions, and potential responses (GAO-11-71). Retrieved from http://www.gao.gov/assets/330/322208.pdf

Vaughn, N. & Lenton, T. (2011). A review of climate geoengineering proposals. *Climatic Change, 109*(3), 745–790.

Victor, D. G., Morgan, G. M., Apt, J., Steinbrunner, J., & Ricke, K. (2009). The geoengineering option: A last resort against global warming? *Foreign Affairs, 88* (64), 64–76.

Vitt, P., Havens, K., Kramer, A. T., Sollenberger, D. & Yates, E. (2010). Assisted migration of plants: Changes in latitudes, changes in attitudes. *Biological Conservation, 143,* 18–27.

Wallace, M. I. (2011). *Green Christianity.* Minneapolis, MN: Fortress.

Walmsley, H. L. (2009). Mad scientists bend the frame of biobank governance in British Columbia. *Journal of Public Deliberation, 5*(1), 1–26.

Walther, G. R., Post, E., Convey, P., Menzel, A., Parmeson, C., Beebee T. J. C., Bairlein, F. (2002). Ecological responses to recent climate change. *Nature, 416,* 389–395.

Warner, K., Ehrhart, C., de Sherbinin, A., Adamo, S., & Chai–Onn, T. (2009). *In search of shelter: Mapping the effects of climate change on human migration and displacement.* Retrieved from http://www.care.org/getinvolved/advocacy/pdfs/Migration_Report.pdf

WBGU (German Advisory Council on Global Change). (2011). *World in Transition. A social contract for sustainability*. Retrieved from http://www.wbgu.de/fileadmin/templates/dateien/veroeffentlichungen/hauptgutachten/jg2011/wbgu_jg2011_kurz_en.pdf
Webber, M. (2011, December 29). How to make the food system more energy efficient. *Scientific American, 306*, 74–79.
Weber, E. U. (2006). Experience–based and description–based perceptions of long-term risk: Why global warming does not scare us (yet). *Climatic Change, 77*, 103–120.
Weinberg, A. M. (1967a). Can technology replace social engineering? *American Behavioral Scientist, 10*, 7–10.
Weinberg, A. M. (1967b). *Reflections on big science*. Cambridge, MA: MIT Press.
West, R. (1987). The feminist–conservative anti–pornography alliance and the 1986 attorney general's commission on pornography report. *Law & Social Inquiry, 12*(4), 681–711.
White, L. (1967). The historical roots of our ecological crisis. *Science, 155*(3767), 1203–1207.
White, R. (1995). *The organic machine: The remaking of the Columbia River*. New York, NY: Hill and Wang.
Whitney, E. (1993). Lynn White, ecotheology, and history. *Environmental Ethics, 15*, 151–69.
Wigley, T. M. L. (2006). A combined mitigation/geoengineering approach to climate stabilization. *Science, 314*, 452–454.
Wilsdon, J. & Willis, R. (2004). *See-through science: Why public engagement needs to move upstream*. London, England: Demos.
Winsberg, E. (2010). *Science in the age of computer simulation*. Chicago, IL: University of Chicago Press.
Wolf, J. & Gjerris, M. (2009). A religious perspective on climate change. *Studia Theologica, 63*, 119–39.
Wolin, S. S. (2010). *Democracy inc.: Managed democracy and the specter of inverted totalitarianism*. Princeton, NJ: Princeton University Press.
Wood, D. J. (2003, August 23). Albert Borgmann on taming technology. *The Christian Century*, 22–25.
Wood, G. (2009, July/August). Re–engineering the earth. *The Atlantic Online*. Retrieved from http://www.theatlantic.com/magazine/archive/2009/07/re-engineering-the-earth/7552/
Woodward, J. (1986). The non–identity problem. *Ethics, 96*(4), 804–831.
Worm, B., Barbier, E. B., Beaumont, N., Duffy, J. E., Folke, C., Halpern, B. S., Watson, R. (2006). Impacts of biodiversity loss on ocean ecosystem services. *Science, 314*, 787–790.
Wynne, B. (2006). Public engagement as a means of restoring public trust in science: Hitting the notes but missing the music. *Community Genetics, 9*, 211–220.
Wynne, B. (2007). Public participation in science and technology: Performing and obscuring a political–conceptual category mistake. *East Asian Science, Technology and Society, 1*, 99–110.
Wynne, B. (2010). Strange weather, again: Climate science as political art. *Theory, Culture, & Society, 27*(2–3), 289–305.
Young, I. M. (2002). *Inclusion and democracy*. New York, NY: Oxford University Press.
Young, I. M. (2006). Responsibility and global justice: A social connection model. *Social Philosophy and Policy, 23*, 102–130.

Index

Academy of Sciences for the Developing world (TWAS), 3
aesthetics, 6
afforestation, 2, 22, 34, 42, 133, 143, 154
Africa, 80, 81, 92, 180
agency-networks, 38, 39, 42n3
agriculture, 22, 32, 39, 51, 80, 102, 139, 193; industrial, 105, 139, 143, 148; small-scale, 143; sustainable, 55–56, 139
air pollution, 122
albedo, 3, 22, 44, 53, 139, 145, 171, 202, 215, 221, 60n4; atmospheric, 171; covering deserts with plastic to increase, 221; crop, 133, 146, 221; roofs, 146
alkalinity, 2
Alliance of Religions and Conservation, 207
aluminum industry, 68
altruist, 16, 18
alzheimer's disease, 163
American Association for the Advancement of Science (AAAS), 2
American Enterprise Institute, 121, 164
American Physical Society, 145
amphibians, 95
Annex, non-Annex countries, 77, 82, 93n1
anthropocene, 234
anthropogenic climate change, 22, 24, 43, 45, 50, 53, 55–58, 77, 97, 101, 103, 103, 104, 135, 176, 182, 185, 202, 215, 93n5; moral significance of, 1; rate of, 103
anti-lock braking systems, 129–130
architect (climate), 228, 232, 234
arctic, 72, 81, 99, 92, 134, 146
Arendt, Hannah, 224, 225
Aristotle, 222; matter and form distinction, 225, 234n2
Arminianism, 213
Army Corps of Engineers, 70
artifact, artificial, 55, 202, 209, 215, 222–228, 233; metastable, 223, 226, 230; stable, 222–223, 226
artificial trees, 2
artisan (climate), 226, 228, 229–231, 232, 233
artist (climate), 228, 231–234
Asia, 22, 26, 80, 81, 92, 180
assessment of geoengineering, 134–142; for social benefits, 133, 147, 148
Augustine, Saint, 213
autonomy, 36, 47–49, 227; of Indigenous peoples, 67
avalanche transceiver, 157, 164
Azerbaijan, 138

back-loaded costs, 51, 53
Bangladesh, 4, 80
beauty, 98, 205
Bergman, Sigurd, 204

255

biochar, 42, 145
biodiversity, 97; loss of, 90, 105; value of, 98, 99–100
biomass, 34, 102
biotechnology, 152, 146, 234; agricultural, 158; medical, 158
Bipartisan Policy Center (BPC), 169
birds, 95
Bonneville Power Administration, 69
Borgmann, Albert, 166, 170
botanical garden, 100
Buber, Martin, 206
Bunzl, Martin, 113, 121, 129
burdens of climate change. *See* harms
buying time, 57, 58, 162, 163, 164, 153, 154, 159, 162–164

Caldeira, Ken, 10n2, 40, 81, 155, 157
Callicott, J. Baird, 99
Canada, 172, 76n1
cancer, 51, 138, 130n4; chemotherapy for, 163
carbon capture and storage (CCS), 102, 106, 109n11. *See also* sequestration of carbon dioxide
carbon dioxide removal (CDR), 2, 22, 28, 34, 38, 41, 42, 43, 145, 151, 154, 159, 177; contrasted with solar radiation management, 2–3, 10n6
carbon footprint, 44
cascading effects, 101
Castoriadis, Cornelius, 227, 232
catastrophe, 18, 55, 59, 67, 70, 71, 107, 134, 154, 190, 196, 218
cathedral building, 230
CDR. *See* carbon dioxide removal
character, 114, 128, 179, 194, 206
China, 82, 165
chlorofluorocarbons (CFCs), 51. *See also* See ozone
Christianity, 9, 204, 208–219; dominion over nature, 208; stewardship of nature, 208, 210, 211
climate change denial, 184
climate debt, 93n2
climate emergency, 3. *See also* catastrophe
cloud whitening, 3, 44, 67, 66, 103, 109n13, 139, 221

co-benefits of geoengineering, 8, 134, 144–146
coercion, 7, 48
Cold War, 68, 70
Columbia River dams, 68–70
co-management regimes, 74
commons, 79; tragedy of, 15
compensation, 84, 85, 87, 156, 157
complexity of the climate system, 1, 4, 90, 106
computer models of global climate, 89–91, 180, 229; as sandpits, 230
consent, 4, 7, 40, 157, 190; free, prior, and informed (FPIC), 72, 75; of indigenous people, 65, 74; integrationist approach to, 72–74, 75; partnership approach to, 75; standards approach to, 71–72, 75
consumer, consumption, 5, 23, 25, 38, 123, 161, 191, 195, 196; wealthy, 6
cooking, home, 196
corals, 95
Corner, Adam, 4–6, 163
corruption, 213; moral, 6, 25, 28, 178, 180, 211
coupled human-environment systems, 138, 139, 140, 143
creation (of artifacts), 221, 222, 227, 228, 231–233
crop productivity, 81, 145; failure of, 142
Crutzen, Paul, 2, 78, 155, 179, 215
cultural traditions, lifeways, 67, 75; impacts of Columbia River dams on, 69–70; impacts of geoengineering on, 71, 74, 90
current generation, 7, 24, 29, 35, 180; the interests of, 51

Dalles dam, 69
debt, 80
Declaration on the Rights of Indigenous Peoples (DRIPS), 75, 76n6
deficit, moral, 8, 79, 83, 83–86, 87, 92
Deleuze, Gilles, 222, 226, 234n2
Descartes, 136, 214; Cartesian binary/dualism, 137, 147
democracy, democratic, 88, 134, 139, 143, 152, 166, 167, 183, 193
desert, 8, 84–87, 88, 90
developed nations, 24, 25, 26, 82, 163

developing nations, 24, 26, 76n2, 78, 79, 81, 83
device paradigm, 166
differential responsibilities, 77, 78, 79, 83. *See also* responsibility
differentiated burdens, 77, 78. *See also* harms
diffuse light, 6
dilemma, 7, 33, 35–42; moral, 18
direct air capture of carbon (DAC), 145. *See also* sequestration of carbon dioxide; scrubbing of carbon dioxide
disaster. *See* catastrophe
discounting, 36, 37
disease, 44, 95
dissipative structures, 223
domination, 7, 43, 47–59
Drengson, Alan, 160
Durkheim, Emile, 19
Dyson, Freeman, 154
dystopia, 218

Earth First!, 160
economic, 116, 143, 144; competitiveness, 144; imperialism, 85
ecosystems, 17, 20, 21, 50, 97, 105, 141, 142; degradation/harm, 23, 29, 79, 106, 134, 144, 173, 174; services, 97, 98, 144
eduction (of artifacts), 222, 226–227, 228, 229–231, 233, 235n3
emergency (climate), 69, 71, 75, 148, 154; response, 157. *See also* catastrophe
emergency medical devices, 157
emerging technology, 1, 158, 169–170
emotional versus rational motivations for behavior, 189–190, 191–192
empathy, 20, 21, 30; empathetic relationships, 20
end of nature, 1, 176, 185. *See also* McKibben
energeia, 224
energy: consumption, 56; efficiency, 38, 193; industry, 139; poverty, 133; security, 144
Endangered Species Act (U.S.), 99
The Enlightenment, 214
enrichment value, 98
Environmental Defense Fund, 3

Environmental Justice and Climate Change Initiative, 73
Erosion, Technology, and Control (ETC) Group, 152, 160
eschatology, 218, 219
ethanol, 125, 139
Ethical, Legal, and Social Implications (ELSI) studies, 91, 93n11
Europe, EU, 31
evapotranspiration, 39, 81
extinction, 17, 59, 97, 101, 103, 107, 96, 155, 108n2, 109n12; rate, 95, 100, 96; historical background, 96, 108n2
extrinsic arguments against geoengineering, 4; versus intrinsic arguments, 4–5, 121; fairness, 15; diachronic, 85–86, 93n6; in distribution of harms, 18, 19, 21. *See also* justice

fallibility, 217, 219
field testing of geoengineering technologies, 7, 34, 37, 88, 89, 151
final value (of species), 96, 98–100
First Nations, 138, 76n1. *See also* Indigenous Peoples
Fisheries, 139; north west Indian, 69–70
flourishing, 192, 197, 198. *See also* the good life
food, 35, 44, 106, 144, 145, 178, 194–196; junk, 196, 198, 196; security, 81, 133, 139, 144–145, 146
foreign aid, 90
form, *eidos* 222–227, 231, 232
fossil fuels, 102, 120, 123, 133, 139, 154, 164, 193, 198; lock-in, 143; taxes on, 198
fracking, hydraulic, 38
framing, 8, 18, 38, 152–167
freedom, 36, 41, 47, 84, 160, 214, 216
free-rider, 117, 126
front-loaded goods, 51
funding of science, 182
future generations, 4, 7, 21, 24, 26, 28, 29, 33, 35, 40, 41, 43, 59, 44, 104, 180; choices available to, 133, 142, 143, 146, 147; moral obligations to, 29, 180, 43–59; rights of, 36, 104
future people. *See* future generations

game theory, 44
Gardiner, Stephen, 1, 6, 25, 28, 66–67, 109n16, 133, 179
Genesis, the book of, 212, 214
geoclique, 162
geoengineering as protecting choice/options, 142
Germany, 82
globalization, 140
global mean surface temperature, 17, 90, 101, 103, 105, 106, 107
global security, 5
global thermostat, the, 4, 23, 28, 29, 211
God/Gods, 127, 204, 209; playing God, 172, 175–177, 216, 234
good life, the, 9, 190, 197, 199
Gore, Albert, 1, 77, 219n2
governance of solar radiation management (SRM), 87, 135, 143, 166, 182–183, 233; bottom-up approach to, 71; structures for, 26, 216
grace, 213–215, 218
Greenland ice sheet, 155
gross domestic product (GDP), 90

Habermas, Jürgen, 88
Hamilton, Clive, 5, 161–162, 176
Hands off Mother Earth (H.O.M.E.), 160
Hanford Engineering Works Project, 68
harms, 23, 32, 79; distribution of, 17, 18, 19, 21, 78, 85, 92, 104, 179–181; (*see also* distributive justice); future, 36, 46; transtemporal balancing of. *See* fairness, diachronic
hedonism, 189
Hegel, Georg, 199
Heidegger, Martin, 225, 229, 232
helmets, athletic, 124, 163
historical responsibility: for greenhouse gas emissions, 18, 77, 83. *See also* intergenerational justice
homo faber 230
hope and fear, 218–219
hubris, 5, 23, 45, 127, 176, 209, 219
human: rights, 30, 35, 75, 104; suffering, 2, 17, 80, 126, 141, 215
humility, 176, 211, 216
hunger, malnutrition, 17, 196, 197. *See* food security

hunter-gatherer, 50, 51
hydrological cycle, hydrology, 4, 22, 81, 231

ignorance, 175, 216
illiteracy, 80
An Inconvenient Truth 1
India, 82
Indigenous peoples, 7, 65–75; sovereignty of, 7, 73, 74, 75
industry, old (oil, steel, coal, chemical), 38
inequity, 18. *See also* injustice
inescapability objection, 50, 52
injustice, 8, 83; climate, 83, 84, 87; compound, 8, 17, 79–83, 87, 91, 92
instrumental value, 96, 97, 98, 100
insure, insurance, 8, 24, 56, 116–118, 153–158, 164. *See also* moral hazard
Intemann, Kristen, 89–91
intentional versus unintentional climate change, 180, 183
intercontinental ballistic missiles, 126
Intergovernmental Panel on Climate Change (IPCC), 78, 219; assessment reports, 78, 80, 90, 95, 101, 135, 151; Special Report on Emissions Scenarios (SRES), 140
international cooperation, 165, 166
International Indigenous Peoples Forum on Climate Change, 73
internet, 197
intrinsic argument against geoengineering, 5, 6
intrinsic value. *See* final value
irreversibility, 6
Israel, Jonathan, 191, 198

Jamieson, Dale, 6, 175, 183, 184, 217
Jerome, Saint, 213
just, justice, 16, 18, 19, 21, 23, 88, 170; distributive, 32, 43, 46, 85, 142, 179–181, 185, 186, 221; global, 30, 79, 144, 183, 190, 193, 198; intergenerational, 17, 28, 43, 46, 46–47, 58, 56, 142, 181, 186; procedural, 4, 26, 65. *See also* fairness

Katrina, Hurricane, 80
Katz, Eric, 160

Keith, David, 3, 71–72, 113, 126, 162, 205
kidney dialysis machine, 157
kinesis 223
Kyoto Protocol, 193

Lagrange Point, 44
land: tenure, 133; use, 122, 144. *See also* agriculture
laziness, 84, 176
legal liability, 180
less-developed nation. *See* developing nation
lesser of two evils, 40, 69; characterization of geoengineering as, 6, 31, 58, 65, 67–68, 75, 153, 215
lock-in, 28, 56, 125, 138, 142–144; agricultural, 143, 146; technological, 27, 29, 88, 89, 142
Lovelock, James, 212, 230

malnutrition. *See* hunger
marginalized people, 7, 89; perspective of, 90. *See also* Indigenous people, developing nations, poor nations
market failure, 116, 117, 123
Marland, Gregg, 2
material device, 190
McClintock, Barbara, 230
McKibben, Bill, 1, 77, 176
McKinsey Report, 194–195
McPhee, John, 230
medical care, 139; economics of, 116
methane, 101
Mexico, 139
migration, 84; human, national policy towards, 84, 87; species, 17, 95
military industrial complex, 38
military use of SRM, 124
mitigation, 25, 26, 29, 36, 38, 39, 40, 41, 34, 36, 37, 40, 41, 44, 104, 107, 108, 145, 153, 154, 184, 229; and adaptation, 18, 26, 38, 39, 44, 50, 53, 55, 56–59, 156; geoengineering as a distraction from, 24, 28, 37, 38, 41, 113, 126, 136, 164; 9.38; (*see also* moral hazard); with SRM as augmentation, 162, 175, 179
momentum. *See* lock-in, technological
monsoon, 22, 26, 39, 81

Montreal Protocol, 26
moral climate, 16, 17, 31
moral community, 15, 29, 31
moral deficit, 79, 83, 84, 92
moral hazard, 5, 7, 8, 23, 24, 38, 79, 107, 113–130, 142, 156, 186; ambiguity of, 119–122, 129; efficiency view, 117, 118, 119, 122, 123–125, 127, 128; responsibility view, 117, 118, 119, 122, 125–127, 128; as a revenge effect, 164; (*see also* revenge effect); vice view, 117, 118, 119, 122, 127, 128; vagueness of, 114–115, 119, 122–129
moral repugnance, 33, 35, 41
moral solidarity, 16, 17, 19, 24
moratorium: ongeoengineering research, 7, 42, 151; on geoengineering deployment, 201, 216

nanotechnology, 158
National Congress of American Indians, 73
national parks, 72
natural disasters, 30
Natural Environment Research Council (NERC), 114, 135
natural value, 107, 108n3, 185. *See also* final value
nature, natural, naturalness, 99, 206; contrasted with artificial, 202; domination of, 23; obligations to, 3, 10, 46; natural process, 202, 97, 185; relationship with, 43, 45
negative emissions technologies (NET), 145. *See also* carbon dioxide removal
Netherlands Environmental Assessment Agency, 135
Niebuhr, Reinhold, 209
Nigerian oil development, 138, 139
non-epistemic values in science, 89–92, 93n8
non-instrumental value. *See* final value

Oak Ridge National Laboratory, 159
obesity, 197
obligations, 2, 8, 17, 19, 21, 35, 53, 55, 83–86; to future generations, 43, 47, 53; to non-human nature, 45
ocean acidification, 24, 79, 81, 101, 106, 107, 120, 122

ocean fertilization, 2, 22, 102, 145
ontological genesis, 227
Otto, Rudolph, 206
over-consumption, 5. *See also* consumer
ozone, 22, 26, 39, 51, 79, 146, 175

Participatory Technology Assessment (PTA), 92
Parfit, Derek: non-identity problem, 35, 60n5
Parthasarathy, Shobita, 72–74
paternalism, 67
path dependency. *See* lock-in
Pelagius, Pelagianism, 9, 208, 212–216, 219
per capita emissions, 83
permafrost, 101, 142
photovoltaic panels 6
Pidgeon, Nick, 4–6, 163
Mt. Pinatubo, 22
plan B, 8, 119, 121, 153–158. *See also* insurance
plant productivity, 6, 101, 106
Plato, Platonist, 225, 227
poiesis 222, 223, 224, 225
political: challenges, problems, 53, 104, 114, 158, 159, 171, 222; decision-making, 164, 166, 170, 186; process, structure, 4, 34, 38, 56, 74, 75, 71, 125, 133, 143, 151, 152; weakness, vulnerability, 4, 79, 77; will, 44, 58, 127, 166, 205
pollution, 121, 122; control 2
posterity. *See* future generations
poverty, poor, 80, 83, 180, 195, 198; nations, 81, 87, 88, 198, 199. *See also* developing nations
power 3.8 3.14 3.15 4.27: arbitrary deployment, 50, 54; formal, 50–52; imbalance in political, 21, 71, 79, 82; (*see also* political weakness); substantive, 50–52, 54
precautionary approach, 175
precipitation, 90; global warming and, 17, 101; geoengineering and, 4, 22, 39, 79, 81, 101, 106, 180
production (of artifacts), 222, 225, 227, 228, 229, 233

public engagement, 76n5, 87; forums for, 87; lab-scale intervention, 91; normative and substantive justifications for, 88–92; upstream, 91, 183
public perception, 9, 169–186
pyrolysis, 2. *See also* biochar

rate of climatic change, 95, 101, 142
Rawls, Rawlsian, 41, 85, 88
recession of 2007-2009, 196
reciprocity, 15, 31
reckless driving, 130
regulatory frameworks, 104, 125, 126, 127, 135, 195
religion, 9, 109n6; climate change and geoengineering, 201–219
The Renaissance, 227
representation. *See* consent
responsibility, 9, 18, 35, 53, 84, 120, 125, 152, 180, 202, 203, 204, 211, 212, 213, 217; historical, for climate change, 17, 80, 83, 148; in a geoengineered world, 1, 139, 183, 190, 232, 233–234. *See also* moral hazard, responsibility view
restoration, environmental, 10, 34, 42, 97, 144, 145; climate, 231
revenge effects, 163–166, 167
reversibility of SRM, 6, 41, 135
rich countries, 8, 36, 83, 84, 87, 92, 79, 136, 198
rights, 36, 48, 85, 134; of future people, 104, 109n4, 30; (*see also* intergenerational justice); of Indigenous peoples, 75; the philosophical concept of, 85. *See also* human rights
risk, 3, 40, 114, 127, 144, 154, 160, 171, 174, 175, 217; assessment of, 136; versus benefit, 163, 221; exposure to, 116, 122, 123; International Risk Governance Council, 131n13; long-term, 36, 39; management of, 71, 164
Robock, Alan, 6, 210
Rolston, Holmes, III, 99
The Royal Society, 2, 3, 77, 113, 114, 135, 153, 162; 2009 Report, 22, 113, 114, 135, 136, 137, 144, 153, 154, 156, 158, 162, 169

sacred, 201, 211

salvation, 213, 214, 215, 216, 218
scarcity, 17, 36; water, 144, 146
Schleiermacher, Frederich, 205
Schneider, Stephen, 2, 210, 211, 212
scrubbing of carbon dioxide, 58, 102
sea level rise, 17, 44, 80, 155, 212
seed bank, 100
self-interest, 6, 16, 21, 29, 53, 82
self-organizing systems, 223
sequestration of carbon dioxide, 2, 58, 102, 105, 107, 108, 193. *See also* carbon capture and storage (CCS)
Sher, George, 84–86
Shue, Henry, 8, 18, 79, 82
side-effects, 22, 23, 25, 50, 55, 136, 137, 139, 163, 173–175; environmental/ecological, 46, 55, 79, 107, 174; unanticipated, 106, 158, 174
sin, 202, 210, 212, 214; original, 213
single actor frame for climate change, 136, 142
slippery slope, 7, 33, 39, 205
smart phones, 194
social development strategies/opportunities, 133, 135
social problems, 8
social science, 9; contrast with ethics, 169, 171, 186; benefits of dialogue with ethics, 170–171
The Solar Radiation Management Governance Initiative (SRMGI), 3, 76n5, 183; 2011 report, 87, 89, 182
solidarity, 7, 17, 19–32, 23; affective dimensions of, 20; and SRM, 22–29; mechanical, 19, 20; organic, 19. *See also* moral solidarity
soteriology, 216, 217, 218
South America, 81
space mirrors, 3, 44, 53, 103, 139, 142, 217. *See also* sun shields
species, 8, 17; final value of, 8, 99–100, 104, 107, 108; instrumental value of, 98; invasive, 44; keystone, 97
Stirling, Andy, 88–89
storms: intensification of, 15; perfect moral, 1, 6, 43. *See also* Gardiner, Stephen
stratospheric aerosols, 2, 3, 5, 6, 22, 78, 81, 134

subsistence farming, 77, 81. *See also* agriculture
sun shields, 221. *See also* space mirrors
superkilling, 99

tar sands development, 38, 138, 139
technological fix, 5, 8, 18, 32, 133, 153, 158–167, 189–190; lure of, 5, 177–179; pragmatic technical fix, 158, 162–167; public preference for, 163, 178–179; risks associated with, 153, 157, 158, 195; techno-fix, 158, 160–162; polarizing nature of, 161–162; pseudo-fixes, 163, 178
technology, 38, 78, 81, 87–92, 102, 108, 114, 138, 147, 154, 158–167; and class stratification, 138; transfer, 78, 84, 87. *See also* technological fix
Tenner, Edward, 163
termination problem, 28, 33, 39, 40, 42, 53, 55, 106, 142, 143. *See also* lock-in, technological
terrorism, 10, 144
theological anthropology, 217, 218
Thernston, Samuel, 164
Tillich, Paul, 205–206
tipping point, 101, 105
topsoil, 55, 102, 105
Tower of Babel, 9, 208, 209–212, 219
traditional ecological knowledge (TEK), 73
transparency of science, 89, 181
transportation, 32, 56, 124, 193, 195
Trivers, Robert, 191–192
troposphere, 3, 221
trust, 15, 26, 27, 29, 31; of scientists/science, 181–182. *See also* solidarity

uncertainty, 88, 90, 106, 175, 228; about effects of geoengineering, 2, 3, 4, 8, 40, 81, 92, 105, 171, 175; political, 40
unconventional alliances, 184
unilateral, unilateralism: deployment of SRM, 27, 60n4, 79, 82, 124, 165; development of SRM, 26
unintended consequences, 55, 158, 163, 167, 174. *See also* side effects; harms
United Kingdom, 80, 82, 144, 172, 182

United Nations Development Program, 207
United Nations Framework Convention on Biodiversity, 151, 183
United Nations Framework Convention on Climate Change (UNFCCC), 77, 78; Conference of the Parties meetings (COP), 78, 82
United Nations Treaty on Environmental Modification (ENMOD), 5
United States, 38, 82, 83, 165, 166, 172, 190, 193–196; Department of Agriculture, 139; Government Accountability Office (GAO), 127, 135; obstruction on climate change, 31
urban heat island effect, 146
utilitarian, 85, 141, 163

vested interests in SRM, 164
volcanic eruptions, 22, 152, 202
vulnerable, vulnerability, 18, 20, 29, 30, 55, 77, 79, 82, 89–92, 180; economic, 80, 81; geographical, 80, 81; nations, 79–80, 82, 87, 88; political, 82; skewed, 81; of species, 95. *See also* poor nations, developing nations

water conservation, 146
wealthy countries. *See* rich countries
weatherization of rocks, enhanced, 2
Weber, Elke, 25
Weinberg, Alvin, 159, 163, 189, 190–191, 192, 193, 198; advocacy of social engineering, 190
White, Lynn, 160, 204
white roofs, 146, 221
wild nature. *See* wilderness
wilderness, wildness, 10, 99, 100, 176
willingness-to-pay, 93n9
Wood, Lowell, 162
Woodrow Wilson Center, 144

zoos, 99, 100

Contributors

Albert Borgmann is Regents Professor of Philosophy at the University of Montana, Missoula, where he has taught since 1970. His special area is the philosophy of society and culture. Among his publications are *Technology and the Character of Contemporary Life* (University of Chicago Press, 1984), *Crossing the Postmodern Divide* (University of Chicago Press, 1992), *Holding On to Reality: The Nature of Information at the Turn of the Millennium* (University of Chicago Press, 1999), *Power Failure: Christianity in the Culture of Technology* (Brazos Press, 2003), and *Real American Ethics* (University of Chicago Press, 2006).

Holly Jean Buck studies media representations of geoengineering and public participation in geoengineering decision making. She earned her MSc in Human Ecology: Culture, Power, and Sustainability from Lund University in Sweden. She has worked as a geospatial humanitarian analyst for the federal government, a mapping technician for a remote sensing company, a sustainability blogger, and a writing teacher. Her current interests include the geopolitics of climate change, place-based geospatial technologies, and environmentally induced migration.

Wylie Carr is a doctoral student in the College of Forestry and Conservation at the University of Montana. He received his BA in Religious Studies from the University of Virginia and his MS in Resource Conservation from the University of Montana. He is currently working with an interdisciplinary team of ethicists and social scientists at the University of Montana on research examining the social and ethical aspects of geoengineering, with a specific focus on broadening discussions of solar radiation management to include more diverse perspectives.

Forrest Clingerman is Associate Professor of Philosophy and Religion at Ohio Northern University. He is coeditor of *Placing Nature on the Borders of Philosophy, Religion, and Ethics* (Ashgate, 2011) and *Interpreting Nature: The Emerging Field of Environmental Hermeneutics* (Fordham, forthcoming). In addition, he has published a number of articles on environmental theology and philosophy. His research focus is how the idea of place is understood spiritually and ethically.

Maialen Galarraga is a Research Associate within the Department of Sociology at Lancaster University. A philosopher by background, her research interests range from environmental philosophy to the philosophy of technology and continental philosophy. Currently she is part of an interdisciplinary project on climate engineering (Integrated Assessment of Geoengineering Proposals), funded by the EPSRC and NERC, that seeks to produce a policy-relevant assessment of geoengineering proposals.

Benjamin Hale is Assistant Professor in the Philosophy Department and the Environmental Studies Program at the University of Colorado, Boulder. He has published papers in journals such as *The Monist, Metaphilosophy, Public Affairs Quarterly, Environmental Values, Science, Technology, and Human Values*, among others. He is currently coeditor of the journal *Ethics, Policy & Environment* and was for two years the director of the Center for Values and Social Policy at the University of Colorado, Boulder. He is also working on two projects: one related to undoing environmental wrongs and the other seeking a nonconsequentialist argument for environmental responsibility.

Marion Hourdequin is Associate Professor of Philosophy at Colorado College. Marion has written on the foundations of the precautionary principle, the ethics of global climate change, the philosophy of ecological restoration, the evolution of morality, and the nature of moral reasons, with papers in journals such as *Environmental Ethics, Environmental Values, Ethical Theory and Moral Practice*, and *Ethics, Policy, and Environment*.

Ashley Mercer is a doctoral student at the University of Calgary whose work focuses on risk perception and understanding of solar radiation management technologies. Her research examines how hazard information is processed and understood between experts and the public. Her other interests include the influence of information framing on risk-related decision processes and epistemological differences between engineering and science. Her past training includes a Bachelor of Arts and Science (Hons.), McMaster University, and a Masters in Public Administration in Environmental Science and Policy, Columbia University.

Konrad Ott studied philosophy at the University of Frankfurt. Since 1997, he has been a full professor of environmental ethics at the Ernst Moritz Arndt University Greifswald. His fields of research are discourse ethics, environmental ethics, theories of sustainability, and ethical aspects of climate change. From 2000 to 2008, Ott has been a member of the German Environmental Advisory Council which counsels the German government on environmental affairs.

Clare Palmer is Professor of Philosophy at Texas A&M University. She studied at Oxford University, and subsequently taught at Stirling and Lancaster universities in the UK, and at Washington University in St. Louis in the United States. She is the author of three books, most recently *Animal Ethics in Context* (Columbia University Press, 2010) and has edited or coedited a number of volumes including, *Killing Animals*, coedited with other members of the UK Animal Studies Group in 2006; and a five-volume collection *Environmental Philosophy*, coedited with J. Baird Callicott, in 2005. Palmer founded the journal *Worldviews: Environment, Culture, Religion*, and edited it for nine years; and she held the position of president of the International Society of Environmental Ethics from 2007-2010. Her main areas of interest are environmental and animal ethics. She is currently working on a co-authored book, *Companion Animal Ethics*, to be published by Wiley-Routledge.

Christopher J. Preston is Associate Professor of Environmental Ethics at the University of Montana. He is the author of *Saving Creation: Nature and Faith in the Life of Holmes Rolston, III* (Trinity University Press, 2009), *Grounding Knowledge: Environmental Philosophy, Epistemology, and Place* (University of Georgia Press, 2003), a coedited collection of essays titled *Nature, Value, Duty: Life on Earth with Holmes Rolston, III* (Springer, 2007), and a special issue of the journal *Ethics and the Environment* on the "Epistemic Significance of Place." He has written articles for a variety of academic and scholarly publications on environmental ethics and on the ethics of emerging technologies.

Ronald Sandler is Associate Professor of Philosophy in the Department of Philosophy and Religion, Director of the Ethics Institute, a researcher in the Nanotechnology and Society Research Group, and a research associate in the Environmental Justice Research Collaborative at Northeastern University. His primary areas of research are environmental ethics, ethics and technology, and ethical theory. He is author of *Character and Environment: A Virtue-oriented Approach to Environmental Ethics* (Columbia, 2007) and *Nanotechnology: The Social and Ethical Issues* (Woodrow Wilson Center,

2009), as well as coeditor of *Environmental Virtue Ethics* (Rowman & Littlefield, 2005) and *Environmental Justice and Environmentalism* (MIT, 2007). His current environmental ethics research focuses on the ethics of species preservation, modification, and creation.

Dane Scott is Director of the Mansfield Ethics and Public Affairs Program and Associate Professor of Ethics in the College of Forestry and Conservation at the University of Montana. Scott's publications and primary research interests focus on deliberative theory and public decisions over emerging technologies. He has recently published articles on the ethical obligations of higher education and the climate crisis as well as philosophy of technology and agricultural biotechnology. Scott was PI on a recently completed NSF project, "Debating Science: A New Model for Ethics Education for Science and Engineering Students." He is editor of *Debating Science: Deliberation, Values, and the Common Good* (Prometheus, 2011). Scott is currently PI on a National Science Foundation project, "The Ethics of Geoengineering: Investigating the Moral Challenges of Solar Radiation Management." Scott teaches a course in UM's College of Forestry and Conservation titled, "Climate Change Ethics and Policy."

Patrick Taylor Smith is a doctoral candidate at the University of Washington, Seattle. His research focuses on global justice and institutional responses to international and intergenerational domination. Patrick has an MA in philosophy from the University of Illinois, Urbana-Champaign. He is currently a Northeastern University Graduate Research Fellow (summer 2011) and next year (2011-2012) will have a research position at the Program on Values in Society at the University of Washington, Seattle.

Bronislaw Szerszynski is Senior Lecturer in the Department of Sociology at Lancaster University, UK, where he also works in the Centre for the Study of Environmental Change (CSEC) and the ESRC Centre for the Economic and Social Aspects of Genomics (Cesagen). He is author of *Nature, Technology and the Sacred* (2005), and coeditor of *Risk, Environment and Modernity* (1996), *Re-Ordering Nature* (2003), *Nature Performed* (2003), and "Changing Climates," a special double issue of *Theory Culture and Society* (2010). His current research topics include climate change and geoengineering, the eco-capitalist imaginary, and urban ethical foodscapes.

Kyle Powys Whyte is Assistant Professor of Philosophy at Michigan State University and an affiliated faculty at the Center for the Study of Standards in Society (CS3), the Peace and Justice Studies Specialization, and the American Indian Studies Program. He is an enrolled member of the Citizen Potawatomi Nation in Shawnee, Oklahoma. Dr. Whyte writes on issues in environmental justice, the philosophies of science and technology, and American Indian philosophy. His articles are published in journals such as *Synthese, Global Ethics, Agricultural & Environmental Ethics, Knowledge, Technology & Policy, Ethics, Place & Environment, Continental Philosophy Review, Environmental Philosophy, Philosophy & Technology, Public Integrity,* and *Rural Social Sciences,* and his research has been funded by the National Science Foundation, Spencer Foundation, and National Endowment for the Humanities. He is a member of the American Philosophical Association Committee on Public Philosophy, the Michigan Environmental and Natural Resources Governance Program, and the Michigan Environmental Justice Working Group.